TECHNICAL
REPORT

# The Industrial Base for Carbon Dioxide Storage

## Status and Prospects

David S. Ortiz, Constantine Samaras,
Edmundo Molina-Perez

Sponsored by the National Energy Technology Laboratory

Environment, Energy, and Economic Development Program

The research reported in this report was sponsored by the National Energy Technology Laboratory and conducted in the Environment, Energy, and Economic Development Program within RAND Justice, Infrastructure, and Environment.

**Library of Congress Cataloging-in-Publication Data** is available for this publication.

ISBN: 978-0-8330-7867-4



Published 2013 by the RAND Corporation
1776 Main Street, P.O. Box 2138, Santa Monica, CA 90407-2138
1200 South Hayes Street, Arlington, VA 22202-5050
4570 Fifth Avenue, Suite 600, Pittsburgh, PA 15213-2665
RAND URL: http://www.rand.org/
To order RAND documents or to obtain additional information, contact
Distribution Services: Telephone: (310) 451-7002;
Fax: (310) 451-6915; Email: order@rand.org

## Preface

Carbon capture and storage (CCS) is the process of capturing carbon dioxide ($CO_2$) prior to its emission into the atmosphere, then storing it in geologic formations on a time scale of hundreds to thousands of years. As part of the U.S. Department of Energy's Regional Carbon Sequestration Partnership program, seven large-scale demonstrations for storing $CO_2$ in geologic formations are either being planned or are currently operating; an additional site is being studied in Canada. Since the 1970s, a network of pipelines has been constructed to transport $CO_2$ for the purpose of using it for enhanced oil recovery (EOR) operations—in which $CO_2$ is injected into a depleted oil field to liberate more oil from the reservoir. Most of the $CO_2$ supplied for EOR operations and demonstration of geologic storage comes from natural reservoirs. If policies mandating the reduction of $CO_2$ emissions were to be enacted, $CO_2$ may be captured from industrial facilities and power plants as one strategy for compliance. In this case, there would be a need to increase the use of $CO_2$ for EOR and geologic storage.

The National Energy Technology Laboratory (NETL) asked RAND to assess the U.S. industrial base supporting transportation and injection of $CO_2$ for EOR and geologic storage. NETL asked RAND to identify and quantify the activities, equipment, and labor required (1) to transport $CO_2$ from a power plant or other source to an injection site, (2) to engage in EOR by $CO_2$ flooding, and (3) to store $CO_2$ permanently in a geologic formation. RAND was also asked to identify those parts of the industrial base pertaining to using and storing $CO_2$ that are shared with exploration, extraction, and transportation of oil and gas and those that are unique to carbon storage and $CO_2$–EOR operations.

This report documents the results of the analysis. It should be of interest to policymakers assessing the implications of policies that would require reducing emissions of $CO_2$ from stationary sources, leading to increased availability of captured $CO_2$ for transport or storage. It should also be of interest to NETL technology managers and participants in the $CO_2$ sequestration program. This report builds on prior RAND research on energy and industrial bases:

- Constantine Samaras, Jeffrey A. Drezner, Henry H. Willis, and Evan Bloom, *Characterizing the U.S. Industrial Base for Coal-Fired Electricity,* Santa Monica, Calif.: RAND Corporation, MG-1147-NETL, 2011.
- Somi Seong, Obaid Younossi, Benjamin W. Goldsmith, Thomas Lang, and Michael J. Neumann, *Titanium: Industrial Base, Price Trends, and Technology Initiatives,* Santa Monica, Calif.: RAND Corporation, MG-789-AF, 2009.

## The RAND Environment, Energy, and Economic Development Program

The research reported here was conducted in the RAND Environment, Energy, and Economic Development Program, a program of RAND Justice, Infrastructure, and Environment. RAND Justice, Infrastructure, and Environment provides insights and solutions to public- and private-sector decisionmakers across numerous domains, including criminal and civil justice; public safety; environmental and natural resources policy; energy, transportation, communications, and other infrastructure; and homeland security. RAND Justice, Infrastructure, and Environment studies are coordinated through four programs—the Institute for Civil Justice; the Safety and Justice Program; the Environment, Energy, and Economic Development Program; and the Transportation, Space, and Technology Program—and the Homeland Security and Defense Center, run jointly with the RAND National Security Research Division. The Environment, Energy, and Economic Development Program research portfolio addresses environmental quality and regulation, water and energy resources and systems, climate, natural hazards and disasters, and economic development, both domestically and internationally. Environment, Energy, and Economic Development Program research is conducted for government, foundations, and the private sector.

Questions or comments about this report should be sent to the project leader, David Ortiz (David_Ortiz@rand.org). For more information about the Energy, Environment, and Economic Development Program, see http://www.rand.org/energy or contact the director at eeed@rand.org.

# Contents

# Figures

# Tables

# Summary

Carbon capture and storage (CCS) is the process of capturing carbon dioxide ($CO_2$) prior to its being emitted into the atmosphere, then either using it in a commercial application or storing it in geologic formations for hundreds to thousands of years. CCS is one means of reducing anthropogenic emissions of $CO_2$ into the atmosphere.

The oil industry already uses $CO_2$ for enhanced oil recovery (EOR) operations, in which $CO_2$ is injected into a depleted oil field to liberate more oil from the reservoir. Many of the systems needed to expand or make possible CCS are in commercial use or are in advanced development and demonstration. The U.S. Department of Energy (DOE) directs a research program to develop and commercialize technologies for the cost-effective capture of $CO_2$ from major sources and for geologic storage. As part of the DOE's Regional Carbon Sequestration Partnership (RCSP) program, seven large-scale demonstrations for storing $CO_2$ in geologic formations are either being planned or are under way. Since the 1970s, a network of pipelines has been constructed to transport $CO_2$ for EOR operations. Currently, most of the $CO_2$ supplied for EOR operations comes from natural reservoirs. If policies mandating the reduction of emissions of $CO_2$ from industrial and power plants were to be enacted, $CO_2$ could be captured from these sources. More EOR or geologic storage would be needed to accept the $CO_2$.

If such a policy were to be enacted, how quickly could the industrial base supporting the transportation and sequestration of $CO_2$ be expanded? To answer this question, the National Energy Technology Laboratory (NETL) asked RAND to assess the industrial base for transportation and injection for $CO_2$–EOR and geologic storage. NETL asked RAND to identify and quantify the activities, equipment, and labor required for the following:

- to transport $CO_2$ from a power plant or other source to an injection site
- to engage in EOR by $CO_2$ flooding
- to permanently store $CO_2$ in a geologic formation.

RAND was also asked to identify parts of the industrial base related to utilizing and sequestering $CO_2$ that have already been developed and are currently utilized by the oil and gas industry, as well as those that are unique to carbon storage and EOR operations. In this analysis we did not evaluate the capabilities of the industrial base to capture $CO_2$; this decision was made to limit the scope of the study so that the analysis could focus on the activities supporting transportation for EOR, and storage of $CO_2$.

## Approach

The industrial base supporting $CO_2$ storage is the collection of capabilities—including equipment, productive capacity, expertise, and labor—that support the development and deployment of $CO_2$ pipelines, EOR operations, and geologic storage of $CO_2$. In the United States, there are already robust industries supporting the manufacture of pipeline components and the construction of pipelines, as well as an oil and gas industry actively engaged in EOR operations, and other outfits capable of developing $CO_2$ storage sites. Determining the capabilities of the U.S. industrial base supporting $CO_2$ transport and storage specifically required that we perform three analytical tasks.

### Define the Activities That Compose the $CO_2$ Storage Industrial Base

To disaggregate the $CO_2$ storage industrial base from related industrial bases supporting natural gas pipelines and oil and gas development, we first identified the activities that specifically support $CO_2$ storage. These activities fall into three areas:

- the design, construction, and operation of $CO_2$ pipelines[1]
- $CO_2$–EOR operations
- geologic storage.

Once these activities were defined, we determined if they were unique to $CO_2$ storage or employed in other sectors, particularly the oil and gas sector. For example, while there are specific requirements for the construction of $CO_2$ injection wells for geologic storage, the techniques for drilling the wells are by and large the same as those used in the oil and gas sector. The activities unique to $CO_2$ storage cannot be fully developed without engaging in actual storage operations. We then quantified the labor and equipment requirements to support each of the activities.

### Generate Scenarios Under Which the $CO_2$ Storage Industrial Base Would Have to Respond

The second step in our analysis was to determine a range of futures bounding the potential demand for $CO_2$ storage. We defined four scenarios resulting from two primary drivers: (a) the existence of a regulatory requirement to reduce emissions of $CO_2$ and a lower relative cost for capture and storage than other technologies for complying with the regulations; and (b) the pace of activity in the oil and gas sector. The first driver determines whether there is a need to develop geologic storage of $CO_2$ on a large scale. The second driver determines the degree to which those developing geologic storage will have to compete for labor, materials, and equipment with the oil and gas sector. These scenarios determined the amount of $CO_2$ that would need to be stored and how much might be consumed for EOR operations. Prior studies conducted by and for NETL were used to bound these scenarios.

### Quantify the Response of the Industrial Base to the Scenarios

The final step in our analysis was to determine how the industrial base supporting $CO_2$ storage would likely respond under the four major scenarios. The responses include estimates of the

---

[1] Prior to transmission by pipeline, captured $CO_2$ must be purified, dehydrated, and compressed into a fluid (ICF International, 2009).

$CO_2$ pipelines that would need to be constructed, the number of EOR and geologic storage sites to be developed, and the amount of key support services that would be needed. Based on these estimates, we were able to determine the ability of the $CO_2$ storage industrial base, in aggregate, to meet potential demands. Using these results, we drew out the major implications for NETL programs.

To support the analytical steps outlined above, we developed a number of detailed cost models using empirically derived data on labor, materials, and capital costs as of 2009, and used these models to generate future cost estimates. We also conducted a set of interviews with industry participants regarding their perceptions of the $CO_2$ storage industrial base, its challenges, and potential.

Our approach relies on two key assumptions. First, we assume that systems to capture $CO_2$ from coal-fired power plants and other stationary sources will be available and deployed in the coming decades, thus providing sufficient $CO_2$ for EOR operations and geologic storage. Whether such systems are actually deployed depends on them being commercially available and the most economic means for achieving compliance with policies and regulations requiring reductions in $CO_2$ emissions. Second, we assume that current efforts to demonstrate the long-term feasibility of geologic storage, monitoring, verification, and accounting of $CO_2$ are successful, thus paving the way for development of this industry.

## Key Findings

### The Activities Supporting the $CO_2$ Storage Industrial Base Are Largely Shared with the Oil and Gas Sector

The $CO_2$ storage industrial base comprises three core activities: transportation of $CO_2$ by pipeline, EOR by $CO_2$ flooding, and geologic storage.

- *Pipelines.* The industrial base used to build and maintain natural gas and petroleum product pipelines is the same industrial base that would be used to build and maintain pipelines to transport $CO_2$. The same steel is used in pipelines in both industries. Pipeline construction techniques, and hence costs, are very similar. The major differences between pipelines used to transport $CO_2$ and natural gas and petroleum products concern the coatings and seals used for $CO_2$, the installation and operation of pumps needed to maintain pressure, and the presence of control valves to allow sections to be isolated for maintenance and to limit releases of $CO_2$ in case of a rupture. According to our analysis, the differences in costs between $CO_2$ pipeline equipment and equipment used in natural gas and petroleum product pipelines do not appreciably affect the ability of the industry to construct $CO_2$ pipelines.
- *$CO_2$–EOR.* Oil recovery by $CO_2$ flooding is already widely deployed commercially by the oil and gas industry. Oil companies survey, prepare sites, drill injection wells, engage in well workovers, and plug wells used in EOR. Activities that are unique to EOR, as opposed to other drilling operations, include storing and injecting $CO_2$. Storage and injection involve receiving $CO_2$ from a bulk pipeline, distributing it throughout the field, injecting it into the field, and separating $CO_2$ from the produced crude oil.
- *Geologic storage.* Many activities supporting geologic storage are shared with the oil and gas sector, including geologic surveying, site preparation, and drilling wells. Injecting

$CO_2$ is an activity shared with $CO_2$–EOR operations. Post-injection monitoring, verification, and accounting (MVA) operations must occur both at $CO_2$–EOR sites intending to demonstrate permanent storage and at geologic storage sites. These activities are unique to carbon storage; the necessary technologies are being demonstrated but have not yet been deployed commercially.

#### $CO_2$–EOR Can Facilitate the Development of Geologic Storage Industrial Capabilities

NETL, through the RCSP, is demonstrating geologic storage of $CO_2$ and developing and testing technologies, systems, and protocols for carrying out MVA activities. From an equipment perspective, injecting $CO_2$ into a deep saline formation is similar to injecting $CO_2$ into a depleted oil reservoir. When $CO_2$–EOR is used for permanent storage, key supporting capabilities are developed. These supporting capabilities include detailed reservoir characterization; operational monitoring of the injected plume of $CO_2$; ensuring that $CO_2$ does not migrate into underground sources of drinking water; and long-term MVA activities.

Additional technologies need to be deployed to support geologic storage of $CO_2$. More subsurface mapping is needed because typically less is known about the geology in the case of geologic storage than for EOR operations, which benefit from detailed knowledge of the production history and geology of the field. Second, tracking and monitoring the $CO_2$ stream during injection will be different in geologic storage applications because there are no producing wells through which oil and $CO_2$ are recovered. Third, the quantity of $CO_2$ that would be injected into a single well is greater than that for a typical EOR injection well. When practiced for the purpose of carbon storage, $CO_2$–EOR advances industrial capabilities for carbon storage, but does not fully develop them.

#### The Carbon Storage Industrial Base Has Largely Demonstrated the Capacity to Meet the Development Needs for EOR and Geologic Storage

Because so much of the industrial base for EOR and $CO_2$ storage is the same or similar to that currently drawn upon for the natural gas and oil industries, we find no major barriers to ramping up operations to support $CO_2$ storage. In particular, we find:

- *The United States has already demonstrated the ability to lay likely needed lengths of pipelines for both EOR and CCS.* To support both EOR and deployment of carbon storage in a timeframe of 2030–2035, a high-end estimate is that up to 32,000 miles of $CO_2$ pipelines would need to be constructed between 2025 to 2035—roughly 3,200 miles per year. The United States has laid similar lengths of natural gas pipeline in the recent past. For example, the U.S. natural gas industry completed 3,600 miles of pipeline in 2008, and 21,000 miles between 2001 and 2010.
- *U.S. industry is likely to be able to hire sufficient workers with the skills needed to lay the potential length of pipeline needed to support both EOR and CCS.* The number of workers in the oil and gas pipeline construction industry grew by about 60 percent from 2005–2008, demonstrating the ability of the industry to quickly recruit and train labor during periods of high demand. In order to meet the upper-bound estimate of $CO_2$ pipeline additions and provide lengths of natural gas pipelines similar to the highest recent annual additions, the capacity of the pipeline construction industry would need to approximately double by 2025. Given the lead time available to build these pipelines

and the likelihood that demand will actually be lower than this upper bound, the U.S. industrial base would likely have sufficient time to expand capacity to meet this demand.

- *We found no constraints on U.S. drilling capacity to expand EOR operations in our high-end EOR scenario.* From 2006 to 2010, an average of seven new EOR projects per year came online. We estimate that a maximum of 120 projects, or approximately 24 per year, would need to come online in the 2030–2035 timeframe. In the context of the overall capabilities of the oil and gas sector, this constitutes a relatively small amount of activity. For example, we estimate the total number of drilling rigs required to support the highest pace of development to be 55, or slightly more than two active rigs per site. Currently, there are almost 2,000 onshore drilling rigs in operation in the United States; the number of rigs required to support EOR development would be a small fraction of the total.
- *We also found no constraints on the availability of drilling rigs or seismic crews to develop geologic storage in our high-end scenario.* Assuming that carbon capture systems are widely deployed soon and that the pace of deployment accelerates, 240 geologic storage sites may need to be opened in the five-year period from 2025–2030, an average of 48 sites per year, to accommodate growing volumes of $CO_2$. We estimate 84 drilling rigs would be required to open 48 sites per year—a small fraction of the total onshore rigs currently available in the United States. We estimate that the number of active seismic survey teams needed to support this scale of development is approximately six, or one-tenth of today's active teams.

## Concluding Thoughts

The NETL RCSPs are in the process of demonstrating geologic storage at commercial scales and in a range of geologies. The partnerships also focus on the development of protocols for monitoring, verification, and accounting for the stored carbon during and after $CO_2$ injection operations. Our analysis indicates that significant expansion of geologic storage capacity is required after 2025 under most scenarios. If we allow several years for permitting and siting of those operations, we conclude that there are approximately ten years before significant injection operations need to begin. Based on the current activity of the partnerships, it appears, from a technical perspective, that the development of geologic storage is on track to meet this goal.

The industrial base for carbon transport and storage could be strained by demand for labor or equipment, much of which is shared with the oil and gas industrial base. During the RCSP demonstrations, NETL has the opportunity to collect data on project activity timelines and overall schedules, the number of qualified bidders, prices for critical equipment, and detailed labor costs. With these compiled data and a comparison with external conditions in the oil and gas market, NETL will be able to ascertain whether the preliminary observed constraints on widespread deployment of carbon transportation and storage are likely to be binding, and determine appropriate and specific R&D strategies or recommended policy responses to alleviate these constraints.

# Acknowledgments

The authors are indebted to many individuals who assisted in the preparation of the document. Over the course of the research, we interviewed several representatives from the oil and gas industry to better understand the details of $CO_2$ transportation and storage. We especially thank Mike Godec of Advanced Resources International, as well as Wayne Rowe and Dwight Peters of Schlumberger Carbon Services for their assistance. Our project sponsors, Timothy Grant and Charles Zelek of NETL, supported the research and provided helpful feedback on early drafts of the document.

This report is much improved due to the helpful comments of the reviewers. Dr. Elizabeth Burton of the West Coast Regional Carbon Sequestration Partnership critically reviewed the document and provided many helpful comments and observations that improved it significantly. At RAND, James Powers also reviewed the document and made many suggestions that we have adopted. We appreciate the support of Keith Crane, our program manager, and our colleagues at RAND. As always, any errors or omissions are the responsibility of the authors.

## Abbreviations

| | |
|---|---|
| BLM | Bureau of Land Management |
| $CO_2$ | carbon dioxide |
| CCS | carbon capture and storage |
| DOE | U.S. Department of Energy |
| DNV | Det Norske Veritas |
| EIA | U.S. Department of Energy, Energy Information Administration |
| EOR | enhanced oil recovery |
| EPA | U.S. Environmental Protection Agency |
| FERC | Federal Energy Regulatory Commission |
| MVA | monitoring, verification, and accounting |
| NAICS | North American Industrial Classification System |
| NETL | U.S. Department of Energy, National Energy Technology Laboratory |
| PHMSA | Pipeline and Hazardous Materials Safety Administration |
| RCSP | U.S. Department of Energy-NETL Regional Carbon Sequestration Partnership Program |
| SCADA | Supervisory Control and Data Acquisition |
| UIC | underground injection control |
| USDW | underground source of drinking water |

CHAPTER ONE

# Introduction and Motivation

## Background

Among the key challenges facing the United States with respect to using fossil energy is the management of emissions of greenhouse gases, of which carbon dioxide ($CO_2$) is the principal component. The U.S. Department of Energy's National Energy Technology Laboratory (NETL) is leading the development of systems for capturing $CO_2$ and is demonstrating the feasibility of permanent geologic storage. Among other activities, NETL sponsors and conducts research to advance technologies, publishes an atlas of potential areas where $CO_2$ can be stored, and sponsors Regional Carbon Sequestration Partnerships (RCSP)—public-private partnerships that are characterizing the storage potential, modeling the mobility and chemistry of $CO_2$ after injection, and performing tests of geologic sequestration. One aspect of carbon capture and storage (CCS) that NETL has not characterized are the logistical, economic, policy, and infrastructure constraints that would limit the rate of storage-site development and the nation's ultimate capacity to store $CO_2$ in a timely manner to meet greenhouse gas mitigation goals.

This study characterizes the industrial base for $CO_2$ storage, including using $CO_2$ for enhanced oil recovery (EOR) operations. The industrial base for CCS is the set of activities carried out by participants in the industry that result in the capture or transport of $CO_2$ for EOR, and permanent geologic storage of $CO_2$. To simplify this analysis, we focus on the downstream activities after the $CO_2$ is captured. Each activity (drilling an injection well, for example) employs labor and equipment, requires time to execute, and has a cost. Many of the activities needed for CCS are used in the oil and gas sector, as are the labor and equipment used in those activities. However, several activities are unique to CCS. Companies would not engage in these activities in the absence of either EOR operations or the demonstration projects supported by the RCSP program.

By characterizing the activities that make up the CCS industrial base, this study assists NETL in its program planning and execution. The study quantifies the potential constraints regarding development of storage sites and the infrastructure needed to support them. NETL may use the results to structure program activities so as to reduce potential strains on the availability of equipment or labor stemming from these constraints. Taking these constraints into account in the development and deployment of CCS, NETL can better estimate the benefits of continued investment in CCS technologies and demonstration. NETL will be able to point to explicit assumptions that drive the constraints, tying estimates of benefits to other energy analyses, such as those published by the U.S. Energy Information Administration (EIA).

The widespread deployment of systems to capture $CO_2$ from stationary sources will not take place in the absence of policies focused on reducing emissions of $CO_2$. The American Clean Energy and Security Act, passed in 2009 by the U.S. House of Representatives but not the U.S. Senate, would have set up a cap-and-trade system to regulate U.S. $CO_2$ emissions. The legislation required an 83 percent reduction in greenhouse gas emissions from 2005 levels by 2050 (U.S. House of Representatives, 2009). Meanwhile, the U.S. Environmental Protection Agency (EPA) has issued performance standards for new power plants that would limit emissions of $CO_2$ (EPA, 2012). Such policies are also being developed at the state level: California is in the process of implementing Assembly Bill 32, also known as the "Global Warming Solutions Act of 2006," which establishes a $CO_2$ cap-and-trade system for the state (California General Assembly, 2006). Should CCS be the most economical compliance strategy, these regulatory efforts might lead to demand for $CO_2$ transport and storage. However, by design, this study considers only the ability of the industrial base to respond, rather than the policy or economic environments that would lead it to respond.

Key questions that our analysis seeks to answer are:

- Can the $CO_2$ storage sector grow rapidly enough to absorb all the $CO_2$ that might become available from deployment of $CO_2$ capture systems? Are available skilled labor and auxiliary services sufficient to support this growth?
- In the absence of a requirement to capture and store $CO_2$, will the expected growth in EOR operations adequately develop the key capabilities needed for geologic storage activities?

## Approach

To perform this analysis, we adapt methods from other RAND industrial base studies (Samaras et al., 2011; Seong et al., 2009). There are three main steps in the analysis.

- *Define activities that make up the $CO_2$ storage industrial base.* For the purpose of this analysis, we consider activities that support the following:
  - the design, construction, and operation of $CO_2$ pipelines
  - EOR operations, including reservoir modeling; field preparation; and $CO_2$ injection, reinjection, and potential storage
  - geologic storage, including reservoir characterization and modeling; injection and monitoring well construction; $CO_2$ injection operations; and long-term monitoring, verification, and accounting (MVA).

  Once these activities are defined, we determine whether they are unique to $CO_2$ storage or shared with other sectors, particularly the oil and gas sector. For example, while there are specific requirements for the construction of $CO_2$ injection wells for geologic storage, the techniques for drilling these wells are very similar to those in use by the oil and gas sector. We also quantify the labor and equipment requirements to support each of the activities.
- *Generate scenarios under which the CCS industrial base would have to respond.* How it responds would depend on the requirements it must fulfill. In this second task, we pose scenarios under which CCS systems may have to be developed and deployed. There are

two primary drivers affecting CCS development and deployment: (1) the existence of a requirement to reduce emissions of $CO_2$, and (2) the pace of activity in the oil and gas sector. The first driver determines whether there is a need to develop geologic storage of $CO_2$ on a large scale. The second driver determines the degree to which those developing geologic storage will have to compete for resources with the oil and gas sector. The resulting four scenarios span the range of possible futures under which CCS systems would develop. In addition to the key drivers, we also note a range of other factors that would affect the activities making up the $CO_2$ storage industrial base.

- *Quantify the response of the industrial base to the development scenarios.* The final step in our analysis is to determine how the industrial base supporting $CO_2$ storage would respond under the four major scenarios. Based on these results, we detail the implications for NETL programs.

To support the analysis above, we conducted a series of interviews with industry participants regarding their perceptions of the $CO_2$ storage industrial base, its challenges, and potential.

Our approach relies on two key assumptions. The first is that systems to capture $CO_2$ from coal-fired power plants and other stationary sources are available and deployed in the coming decades, thus providing sufficient $CO_2$ for EOR operations and geologic storage. Whether such systems are actually deployed depends on them being commercially available and the most economic means for achieving compliance with policies and regulations requiring reductions in $CO_2$ emissions. The second assumption is that current efforts to demonstrate the long-term feasibility of geologic storage and MVA of $CO_2$ are successful, thus paving the way for development of this industry.

## Outline of Report

Chapter Two identifies the activities that make up the $CO_2$ storage industrial base and discusses the status of each. Chapter Three describes the analytical scenarios and other factors we take into account in the quantitative analysis. Chapter Four quantitatively and qualitatively discusses the potential implications for the CCS industrial base of each of the scenarios. Chapter Five presents conclusions and recommendations for NETL. Appendix A provides some additional information regarding occupational codes that support CCS. Appendix B lists major firms composing the $CO_2$ storage industrial base.

CHAPTER TWO

# Defining the Carbon Storage Industrial Base

This chapter defines the core activities of the $CO_2$ storage industrial base and how each of these core activities consists of a set of subactivities. While much of the industrial base supporting $CO_2$ storage is shared with oil and gas exploration and development, a few activities are unique and not exercised by the shared industrial base. We characterize the core activities by the North American Industrial Classification System (NAICS) codes of the industries that carry them out, and provide information on the key equipment, labor skills, and employment in these industries.

## Core Activities of the $CO_2$ Storage Industrial Base

We consider the three primary activities of the industrial base for $CO_2$ storage to be the following:

- pipeline transportation of $CO_2$
- EOR by $CO_2$ flooding
- geologic storage of $CO_2$.

**Figure 2.1**
**Core Activities of the $CO_2$ Storage Industrial Base**

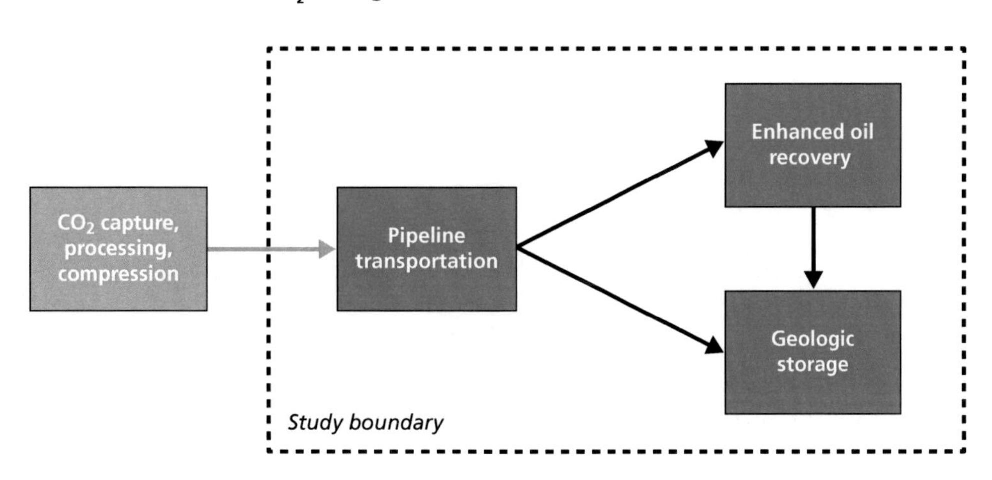


SOURCE: RAND analysis.
RAND TR1300-2.1

The interrelationships among these activities are depicted in Figure 2.1. We omit from this analysis the activities involved in capturing $CO_2$ from a source, processing the $CO_2$ so that it meets specifications for transportation, and compressing it for pipeline transportation. We delineate our study boundary in this way because we are focused on the interaction of downstream activities for $CO_2$ use and storage. We hope to consider the growing industrial base for $CO_2$ capture systems in a future analysis. The remainder of this chapter provides additional details regarding the three core activities.

## Pipeline Transportation of $CO_2$

After $CO_2$ is captured, processed, and compressed, $CO_2$ sources such as power plants and industrial sites are connected by pipelines to $CO_2$ storage sites such as oil fields using EOR or geologic storage sites. Long-distance pipeline transportation of $CO_2$ is a mature technology: The first $CO_2$ pipelines in the United States were installed in the early 1970s (ICF International, 2009). As shown in Figure 2.2, nearly half the existing $CO_2$ pipelines in the United States were constructed during the 1980s, largely driven by new federal tax incentives supporting EOR (Dooley, Dahowski, and Davidson, 2009). More than 4,500 miles of $CO_2$ pipelines have been constructed in the United States, primarily serving $CO_2$–flood EOR sites (Bliss et al., 2010; Pipeline and Hazardous Materials Safety Administration [PHMSA], 2012a). These existing pipelines vary in diameter and capacity: The smallest has a diameter of 4 inches and an estimated maximum flow rate of about 1 million metric tons of $CO_2$ per year, while the largest has a diameter of 30 inches and an estimated maximum flow rate of about 24 million metric tons per year. The existing $CO_2$ pipeline network has a total estimated maxi-

**Figure 2.2**
**U.S. $CO_2$ Pipeline Installation History**

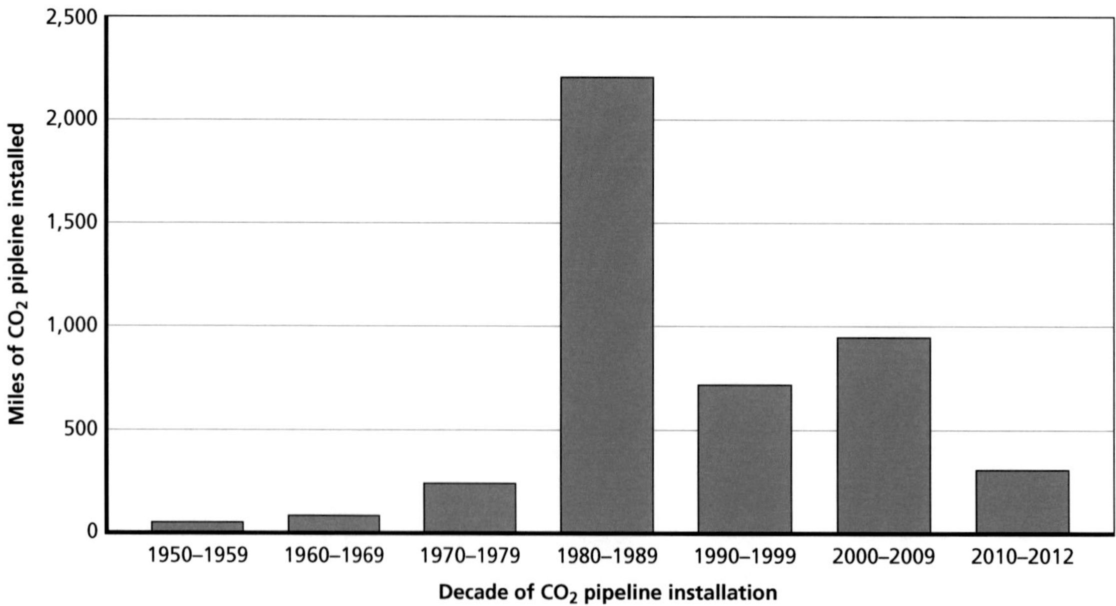


SOURCE: PHMSA, 2012a.
RAND *TR1300-2.2*

mum flow rate of more than 190 million metric tons per year (Bliss et al., 2010). In 2005, the peak year of the last decade, 625 miles of $CO_2$ pipeline were added (PHMSA, 2012b).

The main $CO_2$ pipeline owner-operator firms include Kinder Morgan, Denbury, Oxy, and Exxon. These four firms own and operate more than 60 percent of total U.S. $CO_2$ pipelines as measured by miles and more than 75 percent of total U.S. $CO_2$ pipelines as measured by maximum-flow-rate capacity (Bliss et al., 2010). Despite this concentration, other owner-operators exist. Of the 47 existing U.S. $CO_2$ pipelines listed by Bliss et al. (2010), 18 different firms are listed as owner-operators.

### Similarities and Differences Between $CO_2$ and Natural Gas Pipelines

The design and construction of $CO_2$ pipelines is similar to natural gas pipelines and hence can draw upon the larger, robust natural gas pipeline industry. According to interviewees, most firms that provide $CO_2$ pipeline engineering services are larger firms that also provide oil and gas pipeline engineering services. Hence, the oil and gas pipeline development industry, and its shared capabilities of $CO_2$ pipeline development, provides a strong basis for capabilities in $CO_2$ pipeline engineering and construction should demand for $CO_2$ pipelines increase. Compared with the 4,500 miles of existing U.S. $CO_2$ pipelines, there are more than 300,000 miles of interstate and intrastate natural gas transmission pipelines in the United States, almost all of which are onshore (EIA, undated; PHMSA, 2012b).

But there are several differences between $CO_2$ pipelines and natural gas pipelines relevant to this industrial base analysis. $CO_2$ is transported by pipeline as a dense-phase liquid at pressures up to 2,200 pounds per square inch (Det Norske Veritas [DNV], 2010). Electricity-powered pumping stations maintain the required pressures along the pipeline network (ICF International, 2009). Conversely, natural gas is transported as a gas at 1,000 pounds per square inch; compression stations, rather than pumps, maintain pressure along the system; and compressors along the pipeline often use natural gas as an energy source. The increased pressure requirements of $CO_2$ pipelines necessitate thicker pipes, so thicker steel is used for $CO_2$ pipelines, as shown in Table 2.1. Greater steel requirements increase material costs, transportation costs, and welding costs (ICF International, 2009).

Similar to natural gas pipelines, $CO_2$ pipelines have design requirements limiting the amount of other contaminants that may be transported in the pipeline. These contaminants include water, hydrogen sulfide, sulfur dioxide, and other materials found in natural or anthropogenic $CO_2$ sources. These elements are mostly removed as part of the process of preparing the $CO_2$ prior to insertion in the pipeline. Removing water from the $CO_2$, for example, is critical to maintaining the integrity of the pipeline. The ability of the supplier to remove water and the cost of removal both affect the materials selected for the pipeline (ICF International, 2009). Similar to natural gas pipelines, $CO_2$ pipelines are primarily constructed out of carbon-manganese steel line pipe (ICF International, 2009; DNV, 2010). However, excess water in the system forms carbonic acid, which corrodes the steel. Stainless steel piping or an internal corrosion-resistant coating can be added to carbon manganese steel, but it is generally more economical for long-distance pipelines to remove the water from the $CO_2$ rather than to use more expensive steel alloys or include a pipeline coating (ICF International, 2009). In addition, if a coating becomes detached from the pipeline, it may clog EOR or injection bore holes (DNV, 2010). Removing water from $CO_2$ transported by pipelines is also important to minimize the formation of hydrates—solid, ice-like materials that can plug or damage pipeline components (Element Energy Limited, 2010).

**Table 2.1**
**Thickness and Steel Required for Natural Gas and $CO_2$ Pipelines**

| Type of Pipeline | Pipeline's Outside Diameter (inches) | Final Thickness (inches) | Tons of Steel per Mile |
|---|---|---|---|
| Natural Gas | 12.75 | 0.375 | 130 |
| $CO_2$ | 12.75 | 0.375 | 130 |
| Natural Gas | 16 | 0.375 | 165 |
| $CO_2$ | 16 | 0.419 | 184 |
| Natural Gas | 24 | 0.500 | 330 |
| $CO_2$ | 24 | 0.629 | 413 |
| Natural Gas | 30 | 0.625 | 516 |
| $CO_2$ | 30 | 0.786 | 645 |
| Natural Gas | 36 | 0.750 | 743 |
| $CO_2$ | 36 | 0.943 | 929 |
| Natural Gas | 42 | 0.875 | 1,012 |
| $CO_2$ | 42 | 1.100 | 1,265 |

SOURCE: ICF International, 2009.

For $CO_2$ pipelines, special seals around pipeline valves and fittings are required that are resistant to the physical properties of $CO_2$. Substitution of seals used in natural gas pipelines could lead to seal failure (DNV, 2010). Avoiding leakage is an important planning and design requirement for $CO_2$ pipelines, which require more information than natural gas pipelines about population densities and topography along proposed routes. Because $CO_2$ is heavier than air, it can become concentrated in topographic low points, and it poses a risk to human health at concentrations above approximately 7 percent (DNV, 2010). For this and other reasons, specialized risk-management experience associated with transporting $CO_2$ is necessary for siting and designing $CO_2$ pipelines. To manage and minimize the risks associated with an accidental release of $CO_2$, the block valves, check valves, and vents along the network have to be designed, sited, and installed so as to ensure safety in the case of an accidental release (DNV, 2010). Due to the increased risk of fractures in $CO_2$ pipelines, fracture arrestors are generally sited and installed along the pipeline network to enhance safety (ICF International, 2009; Element Energy Limited, 2010; Gale and Davison, 2004).

### $CO_2$ Pipeline Activities

Figure 2.3 illustrates activities that support the transportation of $CO_2$ by pipeline. Pipelines can be constructed either as connections between a source and a specific use, or as part of a larger pipeline network that connects many sources with many users, such as the existing natural gas pipeline system. Transportation costs on a per-user or per-volume basis could be reduced with a network of large-diameter pipelines (Chandel, Pratson, and Williams, 2010; Kuby, Middleton, and Bielicki, 2011). In the next few decades, however, $CO_2$ pipelines are more likely to be constructed as connecting a single or a few $CO_2$ sources to one or a few CCS sinks, until many more sources of captured anthropogenic $CO_2$ emerge (Bliss et al., 2010).

**Figure 2.3**
**Activities Supporting the $CO_2$ Pipeline Industrial Base**

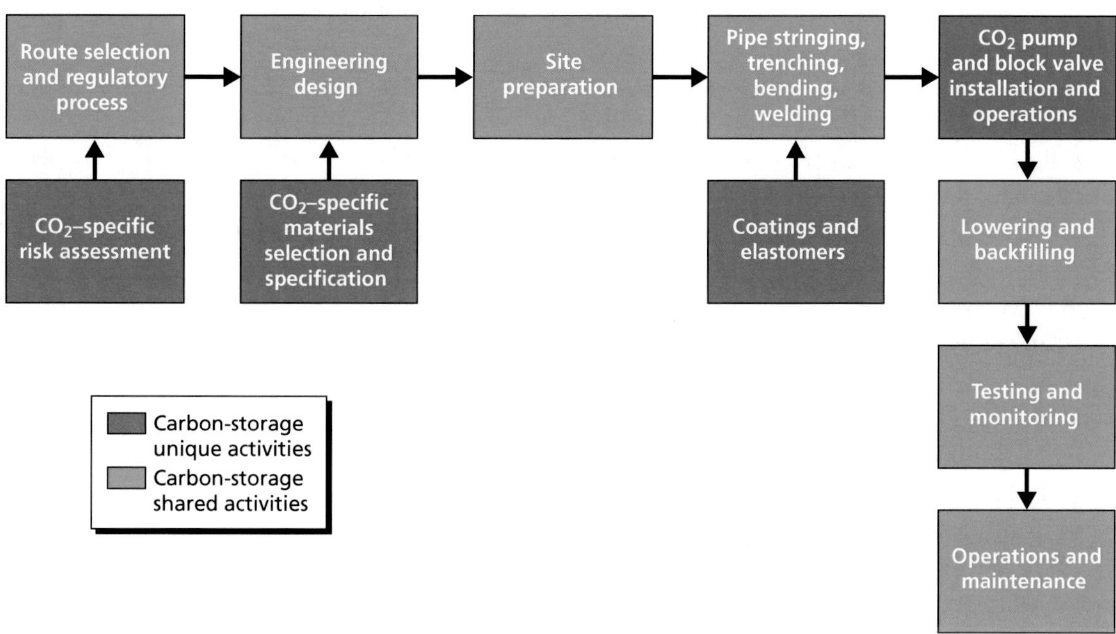


SOURCE: RAND analysis.
RAND *TR1300-2.3*

After a $CO_2$ source and potential $CO_2$ storage project have been identified, a pipeline owner or operator will assess feasible connection routes. Once a preferred route is established, engineering surveys are conducted to determine land right-of-way needs for the pipeline and during construction (Folga, 2007). Land is acquired along the right of way. The owner or operator then applies for appropriate permits from local, state, and federal agencies. The owner or operator also seeks regulatory approval from state regulatory bodies. Historically, unlike natural gas and oil pipelines, which are subject to substantial federal regulation, $CO_2$ pipelines currently require federal regulatory approval only for safety issues and where pipelines cross federal lands (Bliss et al., 2010). For $CO_2$ pipelines installed near population centers, owners or operators take additional risk management and mitigation measures to protect human health in the event of an accidental release (DNV, 2010).

Working with the pipeline owner, engineering design firms specify the pipeline diameter, materials, valve layout, and (if needed) pumping station locations (Element Energy Limited, 2010). These decisions are based on understanding of the $CO_2$ source and use characteristics. Depending on the length of the pipe, volumes of $CO_2$ transported, and route topography, pumping stations along the route may not be needed (Bureau of Land Management [BLM], 2011). Once a design has been accepted and permits acquired, the contractor prepares the construction site, similar to natural gas pipelines. A survey crew marks the centerline of the proposed trench and defines the construction boundaries (Folga, 2007). The land is cleared of vegetation and debris, and graded to provide a level surface for construction. Trenching is completed with either wheel trenching equipment or a backhoe. $CO_2$ pipelines require digging a trench 3 to 4 feet wide and providing 3 to 5 feet of cover above the buried pipe (BLM, 2011). Sections of pipeline up to 80 feet long are shipped by rail to a receiving area, then delivered by truck to the construction site and placed in a continuous line next to the proposed trench (a process termed "stringing"). Using a hydraulic pipe-bending machine, the pipe sections are bent to accommodate the horizontal and vertical direction changes along the route (BLM, 2011).

After bending, the pipe joints are welded together, a process regulated by the Department of Transportation's PHMSA (Code of Federal Regulations [CFR], 2011b). Section 195.222 of the regulation includes language that requires welders to be currently professionally qualified by applicable codes to be eligible to perform pipeline welds.

Welds are then inspected visually by an American Welding Society certified inspector, and radiographic nondestructive testing is performed on a percentage of welds in accordance with PHMSA requirements. The pipelines generally arrive from the manufacturer externally coated with a fusion-bonded epoxy coating to prevent corrosion. An additional coating is applied around joints after welding inspection is complete (BLM, 2011). The location and specification of $CO_2$ pumping stations, block valves, and vents are identified during the engineering design phase; these items are then installed along the pipeline route according to the design. As discussed above, specific $CO_2$–resistant elastomers are applied to all valves and fittings to minimize the potential for accidental leakage.

These welding requirements apply to other hazardous liquid pipelines (such as petroleum); similar requirements apply to natural gas pipelines.[1] A specialized contractor certified by the American Welding Society to conduct radiographic inspection is used for inspection of joint welds (BLM, 2011).

The pipeline is then lowered into place with side-boom tractors. Specialized padding machines create a bedding of soft dirt or other material to support the pipeline in the trench (BLM, 2011). Using a bulldozer, backfiller, or other equipment, the excavated soil is backfilled into the trench and compacted. The pipeline construction is now complete. The pipeline is cleaned by running standard cleaning "pigs" through the pipeline. Prior to operation, hydrostatic pressure testing is conducted to ensure the integrity of the pipeline against leaks (BLM, 2011). During pipeline operations, a Supervisory Control and Data Acquisition (SCADA) control system monitors pipeline pressure and flow to ensure expected operating conditions. Over the lifetime of the pipeline, maintenance consists of minor field repairs due to corrosion. Pipeline sections are replaced if mechanical or other failures occur. These operations and maintenance activities are similar to those undertaken for existing oil and gas pipelines. Similar to other pipelines, cathodic protection is used to minimize pipeline corrosion from the surrounding soils.

### Characteristics

We have mapped the activities depicted in Figure 2.1 to industrial classification codes. We use data corresponding to those codes to characterize the industrial base needed for the design, construction, and maintenance of $CO_2$ pipelines. The data corresponding to NAICS codes relevant to $CO_2$ pipelines is a superset of activities in the sector. For example, these data include U.S. activity for all oil and gas pipeline construction, of which $CO_2$ is a component. The NAICS codes relevant for characterizing the industrial base supporting $CO_2$–EOR are as follows:

---

[1] Natural gas pipeline welders must qualify under the same welding codes, but have slightly different qualification maintenance requirements (CFR, 2011c).

- 237120: Oil and gas pipeline and related structures construction
- 331210: Iron and steel pipe and tube manufacturing from purchased steel
- 332420: Metal tank (heavy gauge) manufacturing
- 333911: Pump and pumping equipment manufacturing
- 333912: Air and gas compressor manufacturing
- 486110: Pipeline transportation of crude oil
- 486210: Pipeline transportation of natural gas
- 486910: Pipeline transportation of refined petroleum products
- 532412: Construction, mining, and forestry machinery and equipment rental and leasing
- 541370: Surveying and mapping (except geophysical) services.

### *Labor*

Figure 2.4 shows the numbers of people employed in sectors supporting pipeline construction in the United States. The legend refers to the NAICS codes listed above. The highest employment in a single sector is that for the construction of oil and gas pipeline and related structures (237210), which peaked at 111,000 in 2008 and declined through 2010 to 92,000. Nongeophysical surveying and mapping (541370) and equipment rental and leasing (532412) employed approximately 60,000 people at their peaks between 2006 and 2008 and dropped after 2008. Employment in all other sectors has held fairly steady over the past decade. The relatively small number of workers directly employed in constructing pipelines is consistent

**Figure 2.4**
**Labor by Sector for Pipeline Industrial Base**

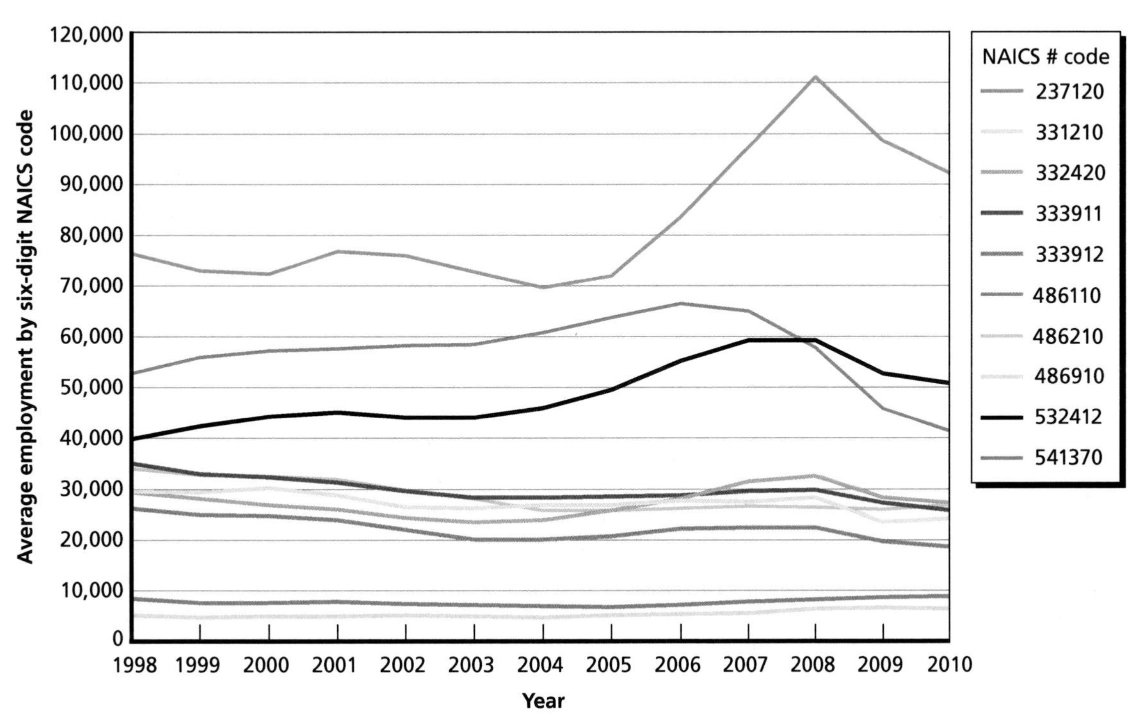


SOURCE: Bureau of Labor Statistics, 2012.
RAND *TR1300-2.4*

with recent experience: The 231-mile Greencore pipeline, under construction in Wyoming, is expected to engage a total of 468 workers (BLM, 2011).

Given the additional welding requirements of $CO_2$ pipelines relative to natural gas pipelines, as well as the welding certifications discussed above, the availability, quality, and cost of welders could be a concern for large $CO_2$ pipeline projects occurring during an environment of rapid oil and gas pipeline development. The process to become a member of the United Association Union of Plumbers, Fitters, Welders and Heating, Ventilation and Air Conditioning Service Techs, which include pipeline welders, involves an apprenticeship that can take up to five years (United Association, 2011). Given potential lead times to train and certify welders, industrial and trade associations could monitor activity and employment so as to flag any potential shortages of welders and other skilled and professional labor. Local training programs and community colleges could be engaged to increase the supply of skilled workers in advance of a surge in $CO_2$ pipeline construction.

Finally, a survey of the natural gas transmission industry found pipeline integrity engineers to be one of the critical and difficult-to-fill positions for the pipeline industry (Interliance Consulting, Inc., 2009). A subcategory of mechanical engineers, pipeline integrity engineers, are responsible for the design and layout of pipeline systems, as well as risk management and safety of the system (Interliance Consulting, Inc., 2009). Engineers in these positions would take into account the risk associated with transporting $CO_2$ when siting and designing pipelines. Given that these positions were deemed difficult to fill by one natural gas industry survey, increasing the supply of knowledgeable pipeline integrity engineers would result from price pressure from the reduced supply in the oil and gas market, and/or targeted university training programs in advance of an anticipated need.

#### *Equipment*

The materials and equipment required for $CO_2$ pipeline construction and operation are similar to those required for oil and gas pipelines. As stated above, the primary material is standard carbon manganese line pipe, which is produced by several manufacturers for the oil and gas industry under American Petroleum Institute Specification 5L (Interstate National Gas Association of America, 2012). Manufactured in steel pipe mills, line pipe can be produced either as welded tubes (generally those with larger diameters) or seamless tubes (generally those with smaller diameters) (Folga, 2007). Pumping stations, pipeline pigs, valves, fittings, cathodic protection and SCADA systems, and other components are manufactured by companies in the larger pipeline industry. As discussed above, specialized $CO_2$–resistant elastomers that are not generally used in the oil and gas industry are used as sealants for $CO_2$ pipeline valves and fittings (Gale and Davison, 2004). Industry interviewees told us that high durometer nitrile elastomers are commonly used in $CO_2$–flood EOR applications, and these are available from major industrial suppliers. Hence, access to these elastomers is unlikely to pose an impediment to wider deployment of $CO_2$ pipelines.

The equipment required to construct pipelines consists largely of construction machinery used in the natural gas and oil pipeline construction industry. A partial list of required equipment, based on a recent Environmental Assessment report for a 231-mile $CO_2$ pipeline that estimated the number of typical machines required to construct the planned pipeline, includes:[2]

---

[2] The Bureau of Land Management provides a full list, including more common construction vehicles (BLM, 2011).

- Dozer with ripper: 1
- Dozer with winch and angle blade: 4
- Sideboom: 8
- Backhoe (3/4 yard): 4
- Ditching machine: 1
- Padding machine: 1
- Bending machine: 1
- Boring machine: 1
- Pipe coating trucks: 1
- Pumps: 3
- Flatbed with winch: 4
- Stringing truck: 6
- Tractor with lowboy: 1
- Welding machine (200-amp, tractor mounted): 1
- Welder's trucks (1 ton): 17

#### *Timelines*

The engineering, permitting, and construction times for $CO_2$ pipelines will be similar to those of oil and natural gas pipelines. For natural gas pipelines, final engineering design can take three to six months (EIA, undated). In advance of construction, the permitting and approvals process can considerably increase project timelines, depending on the necessary permits. If a project could significantly affect federal lands or water bodies, preparation of an Environmental Impact Statement or Environmental Assessment is likely to be required (ICF International, 2009). Moreover, obtaining approvals from landowners regarding rights of way can be very time consuming. Under existing regulations, $CO_2$ and petroleum pipelines do not require federal siting approval from the Federal Energy Regulatory Commission (FERC); natural gas pipelines do (Bliss et al., 2010; ICF International, 2009). FERC siting approval for natural gas projects can be obtained within five to 18 months, with an average of 15 months (EIA, undated). Without the requirement for FERC siting approval, $CO_2$ pipelines could have shorter permitting timelines than natural gas pipelines.

Construction of natural gas pipelines can take six to 18 months (EIA, undated). A typical construction crew can install one mile of natural gas pipeline per day (Folga, 2007). The 231-mile, 20-inch diameter Greencore $CO_2$ pipeline under construction in Wyoming has an estimated construction schedule of two years. However, over this time period, construction is scheduled to occur in only four months each year to avoid extensive impacts on wildlife (BLM, 2011; Davis et al., 2011). Hence, physical construction time of this 231-mile pipeline is approximately eight months, or nearly one mile per day. This demonstrates the potential construction timelines achievable in areas without wildlife or urban constraints.

#### *Costs*

Pipeline costs can be categorized into material, labor, right-of-way, and miscellaneous (design, project management, regulatory fees, and other costs). Material costs, including line pipe and pumping stations, are influenced by commodity prices. The price of carbon steel is a major driver of the cost of $CO_2$ pipelines, accounting for 15 to 35 percent of total pipeline project costs (Bliss et al., 2010). Brown et al. (2011) used natural gas pipeline data filed with FERC to estimate national and regional cost equations for natural gas pipeline construction, categorized by material, labor, right-of-way, and miscellaneous costs, shown in Table 2.2. Material includes line pipe, pipe coatings, cathodic protection, telecommunication equipment, and SCADA systems. Miscellaneous costs include surveying, engineering, supervision, contingencies, telecommunications equipment, freight, taxes, allowances for funds used during construction, administration and overheads, and regulatory filing fees (Liu and Gallagher, 2011).

To estimate construction costs, first estimate the per-mile cost by entering the diameter into the formulas in Table 2.22. Then multiply the result by the distance of the pipeline in miles. As expected, the modeled equations predict that as pipeline diameter increases, material costs become a larger share of total project costs, as shown in Table 2.3 for U.S. natural gas pipelines.

Using the Brown et al. (2011) equations, a 20-inch natural gas pipeline in Wyoming would cost about $54,000 per inch-mile, and a 24-inch pipeline in Texas would cost about $60,000. Reports from recent $CO_2$ pipelines under construction provide some perspective on differences between modeled natural gas pipeline costs and expected $CO_2$ pipeline costs. The 231-mile, 20-inch diameter Greencore $CO_2$ pipeline has an estimated total cost of $275 million to $325 million, or about $60,000 to $70,000 per inch-mile (Fugleberg, 2011). The 320-mile, 24-inch Green pipeline under construction in Louisiana and Texas has an estimated cost of $825 million, or about $110,000 per inch-mile ("American Carbon Capture and Storage Industry Starts Capturing $CO_2$," 2011). Material, labor, right-of-way, and miscellaneous costs will vary by location and by country. Liu and Gallagher (2011) estimated that $CO_2$ pipelines costs in China were about two-thirds of those in the United States and Europe (see, for example, McCoy and Rubin, 2008). Yet, analyses by Schoots et al. (2011) and van der Zwaan et al. (2011) argue that limited cost reductions through learning-by-doing were observed for natural gas and $CO_2$ pipelines, and they do not expect pipeline construction costs to be considerably cheaper with deployment.

## Enhanced Oil Recovery by $CO_2$ Flooding

EOR by $CO_2$ flooding has been practiced in the United States since the early 1970s. A tertiary method for recovering petroleum, it is employed after primary production through conventional

**Table 2.2**
**Derived Cost Equations for Natural Gas Pipelines Per Mile**

| Type of Pipeline | Material | Labor | Right of Way |
|---|---|---|---|
| Region 7 (Texas, Louisiana, Arkansas, Mississippi) | $53418*e^{(0.0799D)}$ | $2.065*50889*e^{(0.0695D)}$ | 2.302*(3480.3*D-15155) |
| Region 8 (Montana, Idaho, Wyoming, Utah, Colorado, New Mexico, Arizona, Nevada) | $53904*e^{(0.0678D)}$ | $2.065*7127.9*D^{1.1641}$ | 2.302*(1112.9*D+19180) |
| U.S. Overall | $63027*e^{(0.0697D)}$ | $2.065*24246*D^{0.9516}$ | 2.302*(1918.8*D+71347) |

SOURCE: Brown et al., 2011.
NOTE: D=diameter of pipeline in inches. Miscellaneous costs are approximately 27 percent of the sum of material, labor, and right-of-way costs.

**Table 2.3**
**Relationships Between Pipeline Diameter and Total Project Cost Category Shares, by Percentage**

| Diameter of Pipeline in Inches | Percentage of Pipeline Modeled Cost | | | |
|---|---|---|---|---|
| | Material | Labor | Right-of-Way | Miscellaneous |
| 16 | 12 | 46 | 15 | 27 |
| 24 | 15 | 46 | 12 | 27 |
| 30 | 18 | 45 | 10 | 27 |
| 36 | 22 | 42 | 9 | 27 |
| 42 | 26 | 39 | 8 | 27 |

SOURCE: RAND analysis using equations from Brown et al. (2011).

wells and enhanced production by flooding the reservoir (typically with water) can no longer produce sufficient crude oil to be profitable. Applications of EOR by $CO_2$ flooding have grown significantly since the technique was first used. Figure 2.5 illustrates the number of active EOR projects by $CO_2$ flooding over time. In 2010, there were 114 active $CO_2$–EOR projects in the United States. This was an increase of 34 projects since 2006. More than half of these projects are in the Permian Basin, which underlies west Texas and New Mexico (Moritis, 2010). The incremental oil production from these projects was 281,000 barrels per day in 2010 (Moritis, 2010).

**Figure 2.5**
**Active Projects of Enhanced Oil Recovery by $CO_2$ Flooding**

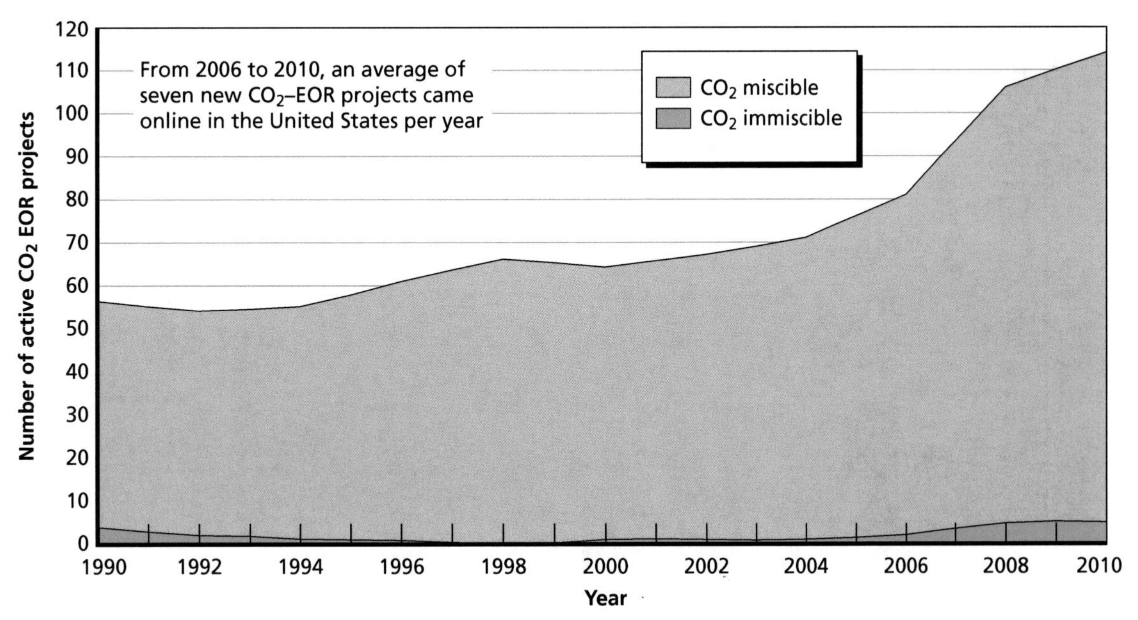


SOURCE: Moritis, 2010.
NOTE: In miscible $CO_2$ floods, the injected $CO_2$ and the oil in the formation mix together, changing the flow properties of the oil and allowing it to be pumped from a producing well. The mixing process is driven by the density of the oil and the high-pressure environment of the reservoir. In immiscible $CO_2$ floods, either oil or reservoir conditions are such that the oil does not dissolve into the $CO_2$, but rather remobilization may still occur in lighter hydrocarbons and by pressure.
RAND *TR1300-2.5*

The three major firms engaging in $CO_2$–EOR operators are Occidental Petroleum, Kinder Morgan, and Denbury Resources (see Figure 2.6). Occidental Petroleum uses EOR most extensively; it has operations in Texas and New Mexico. Total production runs 108,000 barrels per day. Kinder Morgan operates only in Texas and produces 56,000 barrels per day. Denbury Resources operates in Mississippi and Louisiana and produces 34,000 barrels per day. Chevron produces 32,000 barrels per day in Texas, New Mexico, and Colorado (Moritis, 2010).

Figure 2.7 illustrates the activities that support EOR by $CO_2$ flooding. For a given field, the development of an EOR project begins with a geologic survey and the development of a model of the reservoir. Since sites that could use EOR by $CO_2$ flooding have already experienced primary and secondary production, there is likely to be existing data regarding the reservoir, its geologic characteristics, and potential recoverable oil resources. Records of plugged and abandoned wells on the site are most critical from the perspective of $CO_2$–EOR (Davis et al., 2011). These records are usually held by the state agency that oversees oil and gas development. In Texas, this agency is the Texas Railroad Commission. Using these records and geologic surveys, the developer builds a model of the field and drafts a development plan.

Site preparation requirements depend on changes that occurred on the site after primary and secondary oil recovery. The oil fields of west Texas have been under near constant production for many decades, so most are ready for development. Other locations may require acquisi-

**Figure 2.6**
**Production of Petroleum by $CO_2$ Flooding**

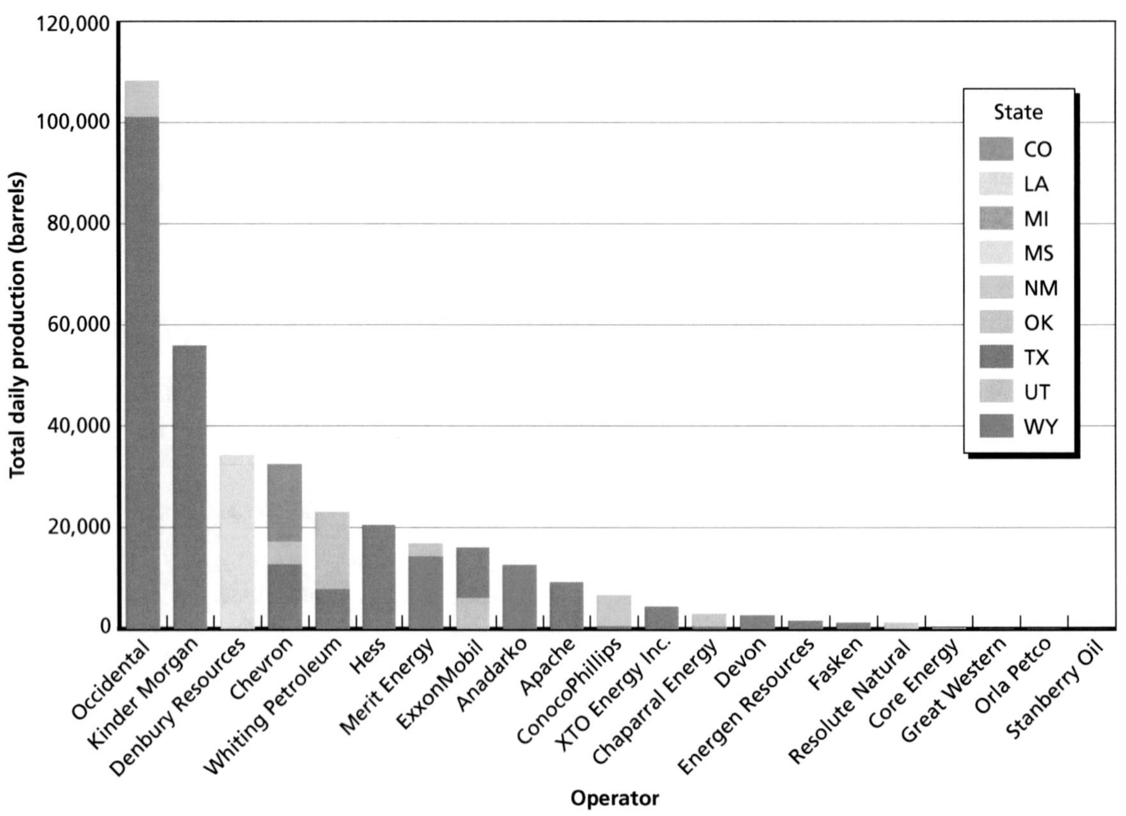


SOURCE: Moritis, 2010.
RAND *TR1300-2.6*

**Figure 2.7**
**Activities Supporting Enhanced Oil Recovery by $CO_2$ Flooding**

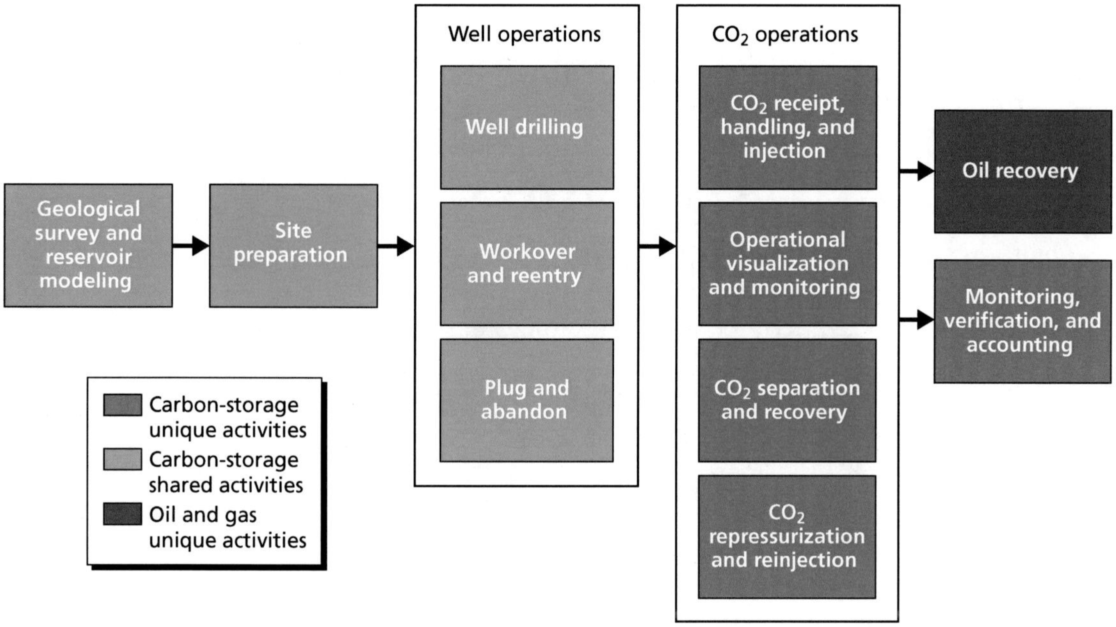

SOURCE: RAND analysis.
RAND *TR1300-2.7*

tion of property and rights to operate. A general timeline for the development of a $CO_2$–EOR site appears in Figure 2.8.

Developing an EOR site requires drilling injection and production wells. To the degree possible, the developer uses existing wells on the site. Because EOR depends on successful injection of liquid-phase $CO_2$ into the reservoir under high pressures, developers find that existing wells require workovers, inspection, and testing. New wells may need to be drilled. Denbury Resources reports that developing one of five blocks of its West Hastings project will require drilling 32 new wells, reentering 30 wells that had been abandoned, and working more than 41 existing wells over a four-year development period (Davis et al., 2011). Denbury is plugging and abandoning 11 wells that are not needed as part of the project. The number of wells varies significantly, depending on the size and connectivity of the oil field and the prior production that occurred there. Occidental Petroleum's Wasson Denver Unit project in Texas

**Figure 2.8**
**General Timeline for Development of an EOR Site**

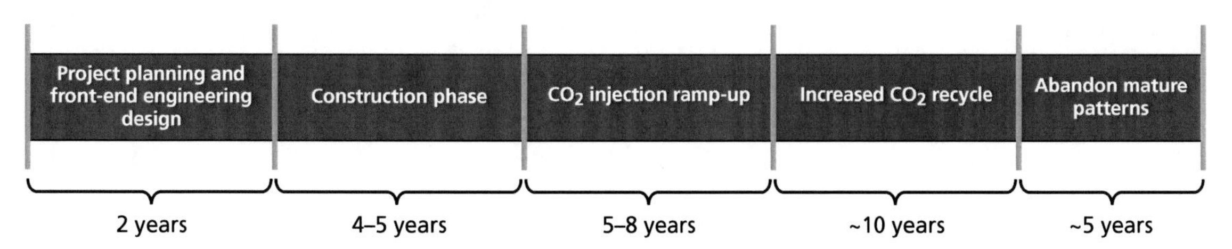


SOURCE: Melzer, 2011.
RAND *TR1300-2.8*

has 580 injecting wells and 1,008 producing wells, whereas Anadarko's Salt Creek Wyoming project has one injecting well and four producing wells (Moritis, 2010). Drilling and initial operations may take several years depending on the site.

Injecting $CO_2$, and the activities required to support injection, set EOR apart from traditional oil and gas operations. These activities appear in the "$CO_2$ operations" box in the block diagram of Figure 2.7. $CO_2$ operations comprise four specific activities:

- receiving the $CO_2$ from a pipeline, distributing it to the injection wells in the field, and injecting specified quantities of $CO_2$ into the formation[3]
- monitoring the movement of the $CO_2$ in the field
- separating $CO_2$ from the recovered petroleum
- repressurizing and reinjecting the $CO_2$ into the field.

As will be discussed later, injecting $CO_2$ into the formation and monitoring its movement is a shared activity with geologic storage of $CO_2$. However, the proposed rules regarding geologic storage wells place more requirements on injection monitoring and in-formation monitoring of $CO_2$ migration than are needed for EOR operations. An integral component of this process is recovering the incremental oil production. Often, the incremental oil will be produced with significant quantities of water, from which it is separated. While injecting $CO_2$ into the reservoir facilitates oil recovery, it is an activity unique to oil and gas operations. Finally, if the $CO_2$ stored as a result of the EOR activities is to meet a compliance requirement, then MVA is also necessary.

The quantities of $CO_2$ that are received and injected can be quite large, especially during the initial years of a $CO_2$ flood operation. Developing and operating a $CO_2$–EOR site requires several decades. During the first several years of operation, produced quantities of oil may be low, while injected quantities of $CO_2$ may be high. As the injected $CO_2$ mixes with the oil in place and the $CO_2$ flows through the field, production increases. As production increases, the amount of $CO_2$ that is recovered from the produced oil grows. Mature EOR projects purchase a small fraction of the $CO_2$ required to support ongoing operations (Melzer, 2011). The Denbury Hastings project has a $CO_2$ capacity of approximately 10 million metric tons per year. This is approximately the amount of $CO_2$ that could be captured from three 500 megawatt coal-fired power plants (U.S. Government Accountability Office, 2010).

### Characteristics

Using the same process we used to identify industrial activities associated with $CO_2$ pipeline construction, we map the activities depicted in Figure 2.7 to industrial classification codes, and use data corresponding to those codes to characterize the industrial base supporting $CO_2$–EOR. The data corresponding to NAICS codes relevant to $CO_2$–EOR is a superset of activities in the sector. These data include U.S. activity for all petroleum production, not only petroleum production associated with $CO_2$–EOR. The NAICS codes relevant for characterizing the industrial base supporting $CO_2$–EOR are:

[3] In practice, water is injected as part of $CO_2$–EOR operations in a method known as "water alternating with gas." The purpose of injecting water is to control the flow of the $CO_2$ through the formation. The amount of water may be tailored to each injection well independently (Advanced Resources International, 2006).

- 211111: Crude petroleum and natural gas extraction
- 213111: Drilling oil and gas wells
- 213112: Support activities for oil and gas operations
- 541360: Geophysical surveying and mapping services
- 333132: Oil and gas field machinery and equipment manufacturing
- 331210: Iron and steel pipe and tube manufacturing from purchased steel
- 332420: Metal tank (heavy gauge) manufacturing
- 333911: Pump and pumping equipment manufacturing
- 333912: Air and gas compressor manufacturing
- 532412: Construction, mining, and forestry machinery and equipment rental and leasing.

The top ten firms participating in each of the NAICS codes above are listed in Appendix B, along with total revenues for 2010.

### *Labor*

The broader industrial base that supports $CO_2$–EOR employed approximately 650,000 people in the United States in 2010. The total employment trends over time for the ten sectors listed above are shown in Figure 2.9. In five of these sectors, employment has remained flat or declined slightly since 1998. These include geophysical surveying and mapping services (541360) and equipment manufacturing industries, including pump and pumping equipment (333911), air and gas compressors (333912), metal tanks (333420), and iron and steel pipe (331210). Three sectors show small increases in total employment: oil and gas field machinery and equipment

**Figure 2.9**
**Total Labor by Sector for the $CO_2$–EOR Industrial Base**

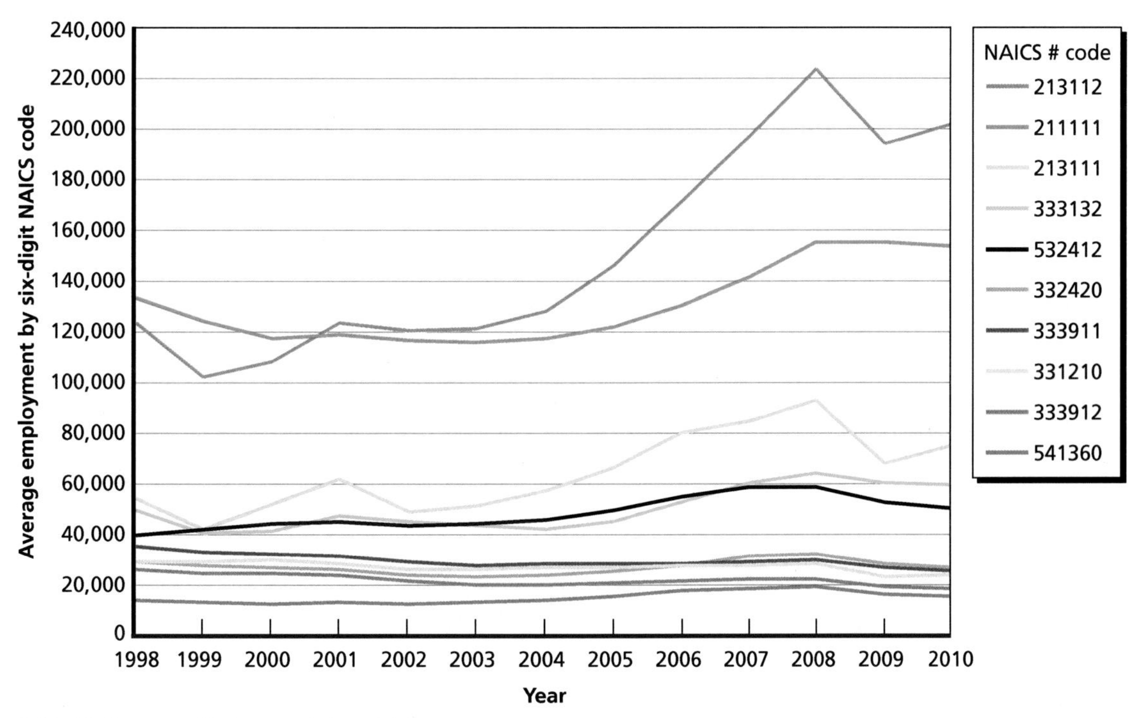


SOURCE: Bureau of Labor Statistics, 2012.
RAND *TR1300-2.9*

manufacturing (333132), geophysical surveying and mapping services (541360), and drilling oil and gas wells (213111). Two sectors show gains of employment greater than 20,000 people from 1998 through 2010, including petroleum and natural gas extraction (211111), and support activities for oil and gas operations (213112).

These trends reflect increased oil and gas activity in the United States; the majority of the employment gains are due to field operations rather than drilling activities. Geophysical surveying services, while in greater demand, have registered a drop in employment in recent years. Note that major oil and gas services companies maintain a geophysical surveying capability so a drop in the number of people employed by specialist firms does not necessarily mean that there has been a drop in the total number of people employed in such activities. The relatively constant employment in related equipment manufacturing, with the exception of oil and gas field machinery, could indicate increases in productivity or be a result of the fact that the oil and gas sector is not the only customer for this equipment; increases in sales to the oil and gas sector may have been offset by declines in demand from other sectors.

Wage trends in these sectors are shown in Figure 2.10. In general, average weekly wages have risen approximately 50 percent from 1998 to 2010. The exception is wages for petroleum and natural gas extraction, which have more than doubled over the period. This rise reflects heavy demand for workers in this sector. Recall from Figure 2.9 that total employment in this sector has not risen significantly.

**Figure 2.10**
**Average Weekly Wages by Sector for the $CO_2$–EOR Industrial Base**

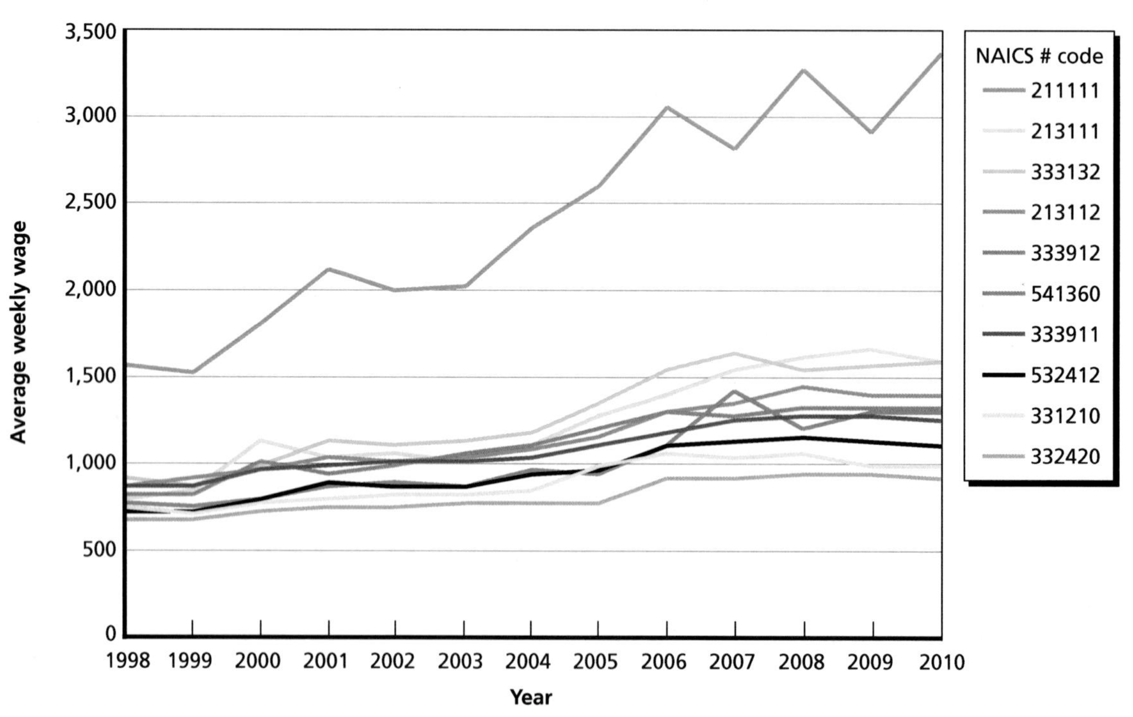


SOURCE: Bureau of Labor Statistics, 2012.
RAND *TR1300-2.10*

### *Equipment*

The equipment to support $CO_2$–EOR operations is similar to that used for conventional oil and gas operations (Davis et al., 2011).[4] Conventional rigs are used to drill new injection and production wells. A workover rig is used to modify, maintain, or plug existing wells on the site. Standard equipment for oil exploration is used to pump the oil liberated by the process.

The infrastructure and equipment that support the injection and recovery of $CO_2$ distinguish EOR operations from conventional oil and gas operations. To support specific operations for receiving, distributing, and separating $CO_2$, a distribution system needs to be built on the site. This system includes a manifold to accept $CO_2$ from a trunk line to allocate it among distribution pipelines at the site; distribution pipelines delivering supercritical $CO_2$ and water to injection wells; pumps, valves, and control systems to manage the flow to the wells; a plant to recover $CO_2$ and water from the oil produced on the site; compressors to convert the recovered $CO_2$ to a supercritical fluid; and heaters to raise the temperature of the $CO_2$ to that of the formation (Davis et al., 2011; Advanced Resources International, 2006.) The wellheads of injection wells need to be designed to support the injection of large quantities of $CO_2$.

### *Timelines*

The development of a field deploying $CO_2$–EOR requires several decades. In general, there are four phases to an EOR operation (Blinclow, 2011; Melzer, 2011):

1. **Field conversion and pilot scale injection.** In this phase of the project, the field is converted or modified to support EOR operations. For example, to develop a single area of the Hastings field, Denbury will drill, modify, or plug wells for four years. Some of these operations can take place while $CO_2$ is being injected (Davis et al., 2011), when $CO_2$ injections are not yet at full scale. This phase of a project typically takes two to six years (Melzer, 2011). This is the time of most activity on the site.
2. **Production ramp-up.** During this phase, which typically lasts four to six years, the operator injects $CO_2$ into the reservoir at full scale. The injected $CO_2$ pressurizes the reservoir and mixes with the oil in the formation, encouraging it to flow. As a result, there is a lag from the time the $CO_2$ is injected to the time the incremental oil is produced. Oil production steadily increases during this phase (Blinclow, 2011; Melzer, 2011).
3. **Production plateau.** During this phase, which lasts between five and seven years, the production of oil from the formation reaches a maximum. The $CO_2$ produced with the oil is separated, repressurized, and reinjected into the formation. Because significant quantities of $CO_2$ are now produced from the oil, purchases of $CO_2$ decline.
4. **Declining production.** As the oil in the formation is extracted, production declines. In a process lasting 20 to 30 years, depending on the size of the field, production falls from the plateau to a level that is unsustainable economically, at which point the field is abandoned. During this period, purchases of $CO_2$ also fall, as sufficient $CO_2$ is recovered from the petroleum to provide for reinjection (Blinclow, 2011; Melzer, 2011).

An operator may repeat the steps listed above in a phased approach to redeveloping a larger oil field, perhaps overlapping the production ramp-up phase with different sections of

---

[4] Interview with Wayne Rowe, Project Manager, Schlumberger Carbon Services, Pittsburgh, Pa., December 2, 2011.

the field. Denbury Resources is taking such an approach to the development of the Hastings field, for example (Davis et al., 2011).

An EOR project in Weyburn, Saskatchewan, has been used as a demonstration for $CO_2$ storage. This project is now entering the MVA phase. The $CO_2$ separated from the produced oil is reinjected into the reservoir and then the injecting and producing wells are sealed, trapping the $CO_2$ in the reservoir. The Weyburn project is estimated to have stored approximately 20 million metric tons of $CO_2$ over its life (Massachusetts Institute of Technology Energy Initiative, 2012).

### *Costs*

Advanced Resources International analyzed the specific cost components related to $CO_2$–EOR for basins throughout the country (Advanced Resources International, 2006). Because of the long lag between initial activities and oil production, Advanced Resources International developed a detailed discounted cash flow model of EOR operations for typical reservoirs. In a recent report, they updated these analyses, including the potential for technological advances to increase production and reduce costs (NETL, 2011a). Typical costs for EOR operations in the Permian basin appear in Figure 2.11. Certain cost components vary according to the price of oil, namely royalties and taxes; in the reference, the price of oil is \$100 per barrel and the total costs are approximately \$44 per barrel.

It is important to put some of the costs into context. Capital costs are averaged over the life and production of the field. However, initial capital costs can be significant: The $CO_2$ recycling plant is a major component of the cost, with installation costs of tens to hundreds of millions of dollars depending on the quantity of gas that needs to be separated and recycled. Well workover and other equipment are leased, so costs associated with this equipment fall into the

**Figure 2.11**
**Typical Current Costs for $CO_2$–EOR in Permian Basin (dollars per barrel produced)**

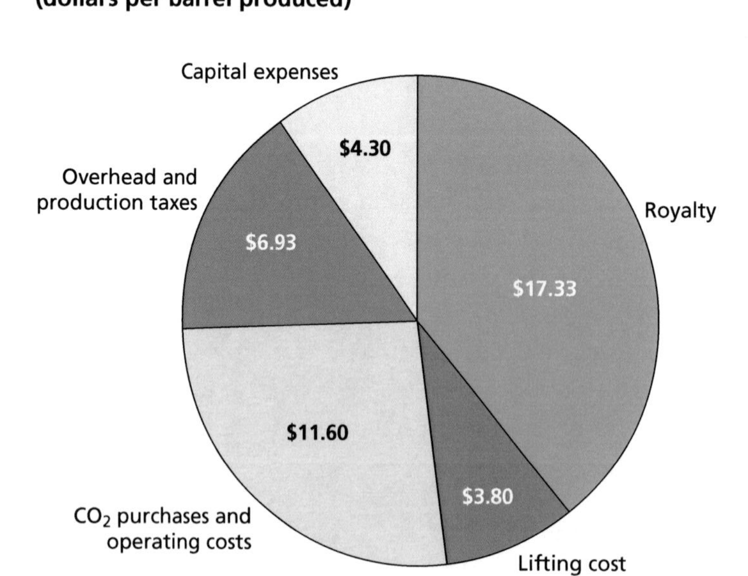


SOURCE: NETL, 2011a.
NOTE: Reference price of oil is \$100 per barrel.
RAND *TR1300-2.11*

category of operating costs, as do labor costs. Lifting costs are the costs of producing oil after wells are drilled and completed, including operations and maintenance of the wells and related equipment, as well as any costs associated with pumping the oil to the surface.

Costs for $CO_2$ and related operating expenses are approximately one quarter of the total cost (NETL, 2011a). Since 2006, average annual $CO_2$ prices have ranged from $29 to $38 per metric ton ($1.50 to $2.00 per thousand cubic feet, quoted at Denver City, Texas) (Wehner, 2011). Continued constraints on supplies from natural sources may lead to higher prices in the future (Wehner, 2011), absent the availability of significant supplies of anthropogenic $CO_2$. Therefore, the proportion of costs attributed to $CO_2$ may rise as well.

## Storage of $CO_2$ in Deep Geologic Formations

The third major activity of the $CO_2$ storage industrial base is permanent storage of $CO_2$ in geologic formations. In the context of this work, we consider storage of $CO_2$ in deep saline formations. Other geologic storage options under investigation include storing $CO_2$ in depleted oil reservoirs as part of EOR operations; storing $CO_2$ in depleted natural gas reservoirs; and storing $CO_2$ in unmineable coal seams (NETL, 2012a).

The core activities supporting geologic storage of $CO_2$ appear in Figure 2.12. Many of these activities are derived from the activities of the oil and gas sector and $CO_2$–EOR operations. Current experience is limited to relatively few domestic and international tests of CCS (as discussed later in this chapter). The geologic surveys and reservoir modeling that are performed in support of geologic storage are very robust (NETL, 2010). In general, developing a new reservoir model of sufficient resolution to support geologic storage operations takes two years.[5] The process is focused on determining the geologic properties of the reservoir and identifying any underground sources of drinking water (USDW) so that $CO_2$ injection operations can be designed to prevent contamination. Models of the reservoir are built and tested against

**Figure 2.12**
**Activities Supporting Geologic Storage of $CO_2$**

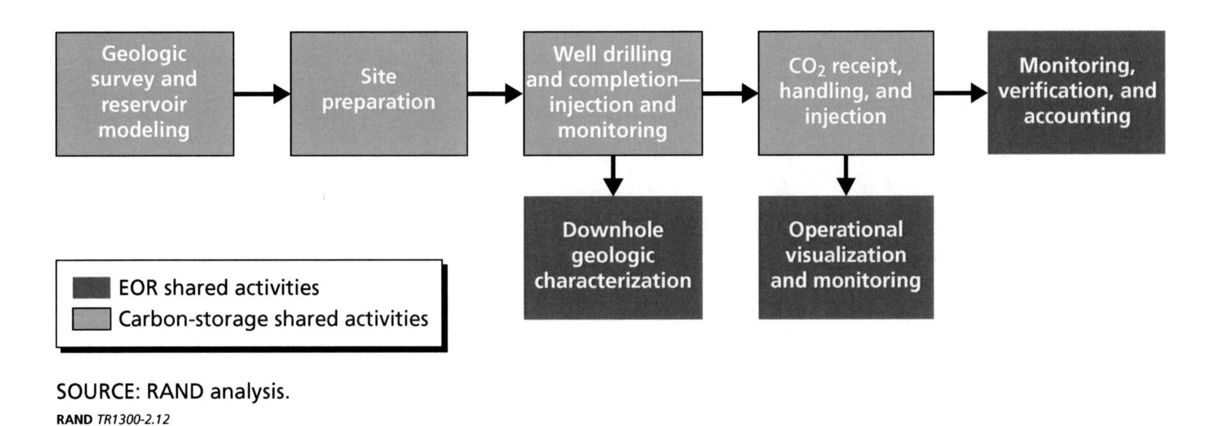


SOURCE: RAND analysis.
RAND *TR1300-2.12*

[5] Rowe interview, 2011.

**Table 2.4**
**Comparison of Requirements for Underground Injection Control Class II and Class VI Wells**

| Item | Class II Well Requirement Applicable to EOR Operations | Class VI Well Requirement Applicable to Geologic Storage |
|---|---|---|
| Site and geologic characterization | Must be sited in formations where there is separation between well and USDW; geologic characterization may be required if area has not yet been characterized; fracture pressure must be estimated. | The area of review[a] must be established; complete characterization of the geologic, hydrogeologic, geochemical, and geomechanical properties of the site is required. Must demonstrate that the proposed volume of $CO_2$ can reasonably be stored in the formation; must be able to predict the movement of $CO_2$ in the formation; core samples must be collected and analyzed; fracture pressure and other properties must be estimated. |
| Well construction | New wells shall be cased and cemented to prevent movement of injected fluids into USDW; casing and cement should be certified for life of well. | Well surface casing must descend through the lowermost USDW and be cemented to the surface; casing and cement should be certified for life of well; at least one long string casing must extend into injection zone and be cemented back to surface; all materials must be compatible with $CO_2$ stream. Mechanical integrity of well must be established. |
| Injection operations | Injection pressure should not exceed fracture pressure. | Injection pressure should not exceed 90 percent of the fracture pressure; the annulus of the tubing must be filled with a non-corrosive fluid and held at positive pressure relative to the $CO_2$. |
| Monitoring and testing | Monthly monitoring of pressure, flow rate, and injected volume; demonstration of mechanical integrity at least every five years; may be monitored on a field basis rather than a well basis. | Reevaluation of the area of review at least once every five years; tracking and monitoring of flow and properties of injected $CO_2$; automatic safety systems are required; monitoring of groundwater; pressure test of well at least every five years. |
| Closure and abandonment | Must be plugged with cement to ensure no migration of fluids into USDW. | Plugging must take place according to a filed plan; monitoring must continue for 50 years after injection. |

SOURCE: CFR, 2011a.

[a] The area of review is the "region surrounding the geologic sequestration project where USDW may be endangered by the injection activity." (CFR, 2011a)

data, refined as needed, and used to construct the site development plan (NETL, 2010). Site preparation activities are the essentially the same as those for oil and gas.

Drilling operations for geologic storage differ somewhat from comparable oil and gas operations. There remains an emphasis on ensuring that USDW are protected. These differences are a result of requirements for wells under the underground injection control (UIC) Class VI designation, which covers $CO_2$ injection wells, whereas Class II wells cover $CO_2$–EOR operations. A comparison of the requirements under each of these regimes appears in Table 2.4. EOR operators who wish to convert their sites for the purpose of geologic storage or to store $CO_2$ as part of a compliance requirement must demonstrate that they satisfy the requirements of Class VI wells. There remain some uncertainties regarding how to account for stored $CO_2$ during EOR operations.

In accordance with the requirements above, the developer drills and completes $CO_2$ injection and monitoring wells (Koperna et al., 2009).[6] The drilling process takes approximately

[6] Rowe interview, 2011.

90 days for a $CO_2$–injection well, compared with approximately 30 days for oil and gas wells of similar depth, because the $CO_2$–injection well must be logged frequently and the operator must conduct down-hole seismic imaging, which requires drilling operations to stop.[7] For example, drilling of the injection well for the Decatur Project took 79 days (NETL, 2010). If geologic storage reservoirs are at depths much greater than oil and gas activity in a particular basin, the size of the rig used for a geologic storage well may be larger than rigs regularly used on comparable oil and gas wells. UIC Class VI requirements stipulate that the surface casing for a geologic-storage well must pass through the deepest USDW. If the base of the USDW is deep relative to the storage horizon, then the mass of the surface casing for a UIC Class VI well may be heavier than that for a typical oil or gas well, requiring a larger rig.[8] Industry interviewees told us that previous $CO_2$–flood EOR projects generally used drilling rigs with a static hook load capacity of 250,000 pounds. Interviewees went on to say that onshore oil and natural gas rigs generally are grouped by static hook capacities of 250,000 pounds and less, 350,000–450,000 pounds, 650,000–800,000 pounds, and greater than 1 million pounds. While there are differences among the well requirements for geologic storage and those for oil and gas operations, geologic storage employs techniques already developed to support oil and gas drilling, so the industrial base is shared.

As noted in Table 2.4, there are some differences in the requirements regarding injecting $CO_2$, specifically with respect to maintaining positive pressure in the annulus—the section of the well surrounding the injection tubing—and monitoring the movement of the $CO_2$ plume. Because the activities are derived from those developed to support EOR operations, we denote them as shared activities in Figure 2.12.

The unique activity supporting geologic storage of $CO_2$ is the long-term MVA of the injection well and the migration of the injected $CO_2$. Operational approaches for MVA activities are currently in development as part of the RSCP administered by NETL. MVA research focuses on four major areas: atmospheric monitoring technologies; remote sensing and near-surface monitoring technologies; subsurface monitoring technologies; and the design of intelligent monitoring systems and protocols (NETL, 2012b). This is a broad range of activities that draw on techniques developed to monitor air quality and perform satellite-based remote sensing, as well as subsurface surveys developed to support oil and gas operations. The goal of the NETL program is to develop a suite of technologies and approaches to MVA that can be tailored to the specific needs of an injection site (NETL, 2012b). We assume that these efforts will be successful and that protocols for MVA will be developed and available as captured $CO_2$ becomes available.

MVA comprises a range of related activities focused on two general tasks: (1) accounting for stored $CO_2$ for the purposes of compliance with emissions reductions; and (2) ensuring that the storage of $CO_2$ does not pose health or environment risks (Morgan and McCoy, 2012). A number of methods and technologies will have to be brought together to perform these tasks, including monitoring the amount of injected $CO_2$, measuring the mass of $CO_2$ in situ, detecting potential leakage through remote sensing, and other methods. Such activities will also have to occur during $CO_2$–EOR if it is performed for the purpose of permanently storing $CO_2$. For example, as part of the development of the Hydrogen Energy California project,

---

[7] Rowe interview, 2011.

[8] Rowe interview, 2011.

which includes integrated $CO_2$–EOR operations, a plan has been drafted and submitted to state officials regarding the monitoring, reporting, and verifying storage of $CO_2$; complying with evolving regulations; and minimizing potential public health impacts (Occidental of Elk Hills, 2010).

The activities of the RCSP are being carried out in three phases. The "characterization phase" collected data on geologic formations and the potential for these formations to store $CO_2$, and developed capabilities for future testing of geologic storage. The "validation phase" comprised many small-scale injections of $CO_2$ into geologic formations—i.e., an injection rate on the order of a few thousand metric tons over a timeframe of days to months (NETL, 2012c). Currently, the RCSP is in the "development phase," which is carrying out eight large-scale, long-term injections of more than 1 million metric tons per year (NETL, 2012d).

Since most of the activities supporting geologic storage are shared either with oil and gas operations or with $CO_2$–EOR, the NAICS codes describing these activities overlap as follows:

- 213111: Drilling oil and gas wells
- 213112: Support activities for oil and gas operations
- 541360: Geophysical surveying and mapping services
- 333132: Oil and gas field machinery and equipment manufacturing
- 331210: Iron and steel pipe and tube manufacturing from purchased steel
- 332420: Metal tank (heavy gauge) manufacturing
- 333911: Pump and pumping equipment manufacturing
- 333912: Air and gas compressor manufacturing
- 532412: Construction, mining, and forestry machinery and equipment rental and leasing.

Because this list of activities excludes only oil and gas extraction (code 211111) when compared with the activities supporting $CO_2$–EOR, we refer the reader to Figures 2.10 and 2.11 to describe the employment and wages prevalent in those activities. The preceding list omits air quality and remote sensing aspects of MVA that are currently under development.

### Characteristics

#### *Labor*

The key difference between labor inputs for $CO_2$–EOR and geologic storage is the above-stated absence of oil and gas extraction as one of the key activities. Otherwise, the characteristics of the labor base supporting geologic storage are similar to those for $CO_2$–EOR. As also noted above, total employment in these sectors has grown principally in oil and gas field services since 1998, and less so in drilling oil and gas wells. As indicated in Figure 2.9, employment in firms specializing in geophysical mapping services has declined slightly, although this fall may not indicate a decrease in U.S. geophysical mapping capabilities because these operations within larger companies are not captured in these data. The geophysical mapping and characterization of the injection site are essential activities for supporting geologic storage.

#### *Equipment*

Given that the core activities supporting geologic storage are similar to those supporting $CO_2$–EOR, the equipment requirements are also similar with a few key exceptions. First, the $CO_2$ receiving and distribution system is simpler than that for an EOR operation because it does not need to include an oil and gas separator. The system in the field for delivering $CO_2$ to the injection well may be simpler because there may only be one or several injection wells as compared to potentially dozens for a large EOR operation. The system may include line heaters and pumps to prepare the $CO_2$ for injection into the geologic formation.

The second principal difference is in the size of the equipment used to drill the injection well. The diameter, thickness, and length of the surface casing are all generally larger than would typically be required for oil and gas wells. As a result, as noted above, if a geologic storage well in a particular region is deeper than typical oil and gas wells, a larger drilling rig may be required than would be available, thus increasing costs. Additionally, to obtain a rig not currently used in a particular region (such as a rig for a geologic storage well), the developer may have to pay to mobilize and move the rig to the site, which adds substantial costs.[9]

#### *Timelines*

It takes several years to develop and begin operations on a geologic storage site. The Gulf Coast Stacked Storage Project at the Cranfield oil field in Mississippi, carried out by the Southeast RCSP, which tested $CO_2$ storage through EOR, began site characterization for its Phase II test in March 2007, completing it in a year (Grant, 2012). Site development occurred concurrently and was complete by July 2008, when injection operations began. The Decatur Project, carried out by the Midwest Geologic Sequestration Consortium, which captured $CO_2$ from an ethanol processing facility to test storage in a saline formation, began site characterization in October 2007 and began injection operations in August 2010 (NETL, undated; Midwest Geologic Sequestration Consortium, 2012; Grant, 2012). The plans for the Big Sky Regional Carbon Sequestration Partnership Phase III geologic storage test call for approximately one year to obtain permits and to comply with the requirements of UIC Class VI wells (CFR, 2011a; Big Sky Carbon Sequestration Partnership, undated). The permits for UIC Class VI injection include those associated with site characterization; well construction, management, and monitoring; and financial responsibility. Other permits may include construction permits for pipelines, land-use permits allowing drilling on the site, and permits guaranteeing access to the subsurface pore space. Prior to the permitting phase, project developers may spend approximately one year gathering and analyzing existing data to support permit applications.[10] Moreover, developers devote resources during this time to engage the public and to address stakeholder concerns.

Another year or more may be spent on risk assessment. Risk assessment is a broad activity including a detailed analysis of technology, operating procedures, and subsurface characteristics. Reservoir characterization is a key component (Big Sky Carbon Sequestration Partnership, undated; NETL, 2011b).[11] The purpose of the reservoir characterization and modeling phase is to collect data for the permitting process and to estimate the storage potential of the reservoir along with the physical and chemical dynamics of injecting $CO_2$ into the formation. It also includes a detailed characterization of the formation's USDW and any faults in the formation that could lead to leakage of injected $CO_2$ from the intended storage site (NETL, 2011b; NETL, 2011c). The techniques that may be used include surface seismic imaging, analysis of core samples, down-hole seismic imaging, wire line logging, and gravity and magnetic

---

[9] Rowe interview, 2011.

[10] Rowe interview, 2011

[11] Rowe interview, 2011.

surveys, among others. Toward the end of the permitting process—or after permits have been approved—the infrastructure is constructed that will be used to transport $CO_2$ to the storage site and distribute it upon arrival. For the Big Sky Carbon Sequestration Partnership, this process is expected to last between six and nine months (Big Sky Carbon Sequestration Partnership, undated).

Once permits are received, $CO_2$ can be injected into the formation. Injection tests may be relatively short, lasting less than a month, as in the case of the 2008 test at the Victor J. Daniel Power Plant in Mississippi (Koperna et al., 2009), or last several years, as was carried out in the Southeast Regional and Midwest tests (Southeast Regional Carbon Sequestration Partnership, undated; NETL, undated; Midwest Geologic Sequestration Consortium, 2012), and is planned for the Big Sky Carbon Sequestration Partnership (Big Sky Carbon Sequestration Partnership, undated). Full-scale $CO_2$ injection operations may last several decades. To date, there have been three large-scale, long-term tests of geologic storage. The one in Weyburn, Saskatchewan, used $CO_2$ delivered by pipeline from a coal-gasification facility in North Dakota for EOR, injecting a total of 18 million metric tons over 10 years (Massachusetts Institute of Technology Energy Initiative, 2012). The project is entering an MVA phase. The Sleipner project, operated by Statoil in the North Sea since 1996, injects approximately 1 million metric tons per year of $CO_2$ separated from natural gas processing into a saline formation. The In Salah project in Algeria, started in 2004, injects $CO_2$ to increase natural-gas recovery (Toman et al., 2008). Both the Sleipner and In Salah projects plan to continue injection of $CO_2$ as part of natural gas recovery operations, and both engage in detailed monitoring of the migration of the injected $CO_2$ (Massachusetts Institute of Technology Energy Initiative, 2012; Toman et al., 2008).

During injection, operations are subject to detailed monitoring. Migration of the injected $CO_2$ in the formation is tracked. These data are used to assist in operations and to modify and improve the existing reservoir model. When injection operations are completed, the injection wells are closed and the project transitions into the post-injection site care phase of the project. This phase involves continued monitoring of the $CO_2$ plume until pressures and $CO_2$ concentrations in the injection zone stabilize. When the governing regulatory authority concludes that the $CO_2$ plume has stabilized, these activities conclude, monitoring wells are closed, and other devices are removed. The site enters the long-term stewardship phase, where there may be little or no active monitoring and the site can be used for other purposes. MVA activities are an integral part of the project from beginning to end (NETL, 2009), and become the primary activities during the closure and post-closure phases of the project. Figure 2.13 presents a generic timeline of these activities, based on the NETL model of geologic storage costs.

#### *Costs*

The costs of geologic storage are not well established. Using data compiled by the EPA (2010) and through the RCSP, NETL has developed a cost-estimation tool for geologic storage (NETL, 2012e). The activities modeled by the tool are those required for carrying out the activities required for UIC Class VI wells (CFR, 2011a). The specific site that is modeled has two injecting wells, two in-formation monitoring wells, and a set of shallower wells for monitoring water supplies. The site characterization activities require three years and include full seismic imaging of the site, core sampling, and aerial imaging. Operations include all activities associated with operating and maintaining the well. Site closure (post-injection site care) includes plugging the wells and 50 years of on-site monitoring. The total real costs of these

**Figure 2.13**
**Timeline for a Geologic Storage Site**

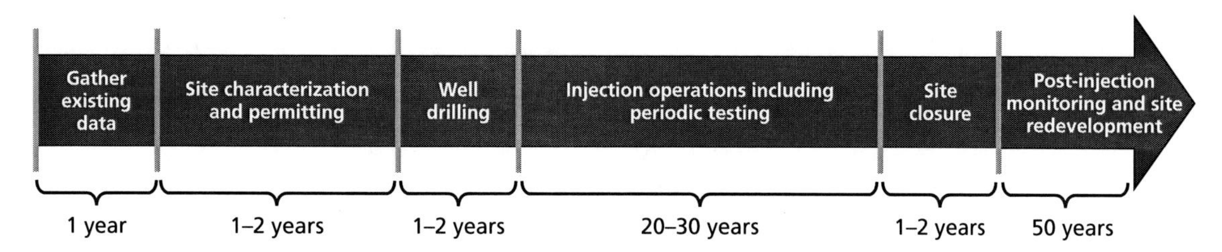


SOURCE: NETL, 2012e; Rowe interview, 2011; Big Sky Sequestration Partnership, undated.
RAND *TR1300-2.13*

activities are estimated to be approximately $300 million, slightly more than three-quarters of which occurs during operations. Figure 2.14 illustrates the total costs of commissioning, operating, and closing the site according to the model.

**Figure 2.14**
**Representative Costs of Geologic Storage of $CO_2$ (millions of dollars)**

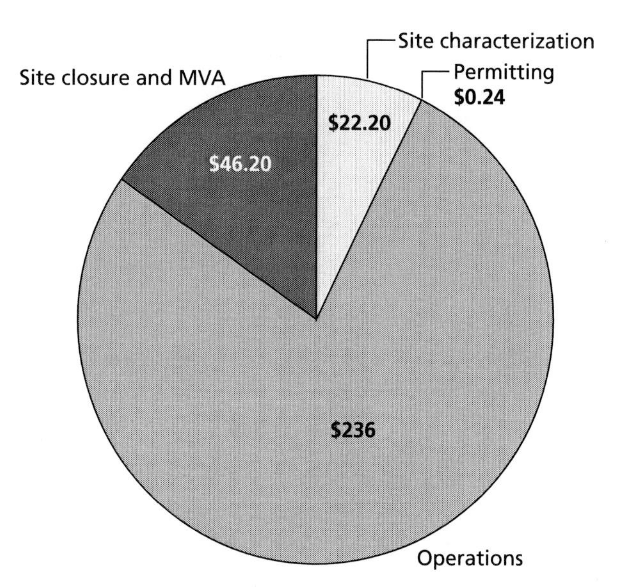


SOURCE: NETL, 2012e.
NOTE: Costs are in real 2010 dollars.
RAND *TR1300-2.14*

CHAPTER THREE

# Development Scenarios for CCS

This chapter details four scenarios under which CCS systems could be deployed, and estimates the number of EOR projects with storage or geologic storage sites that would have to be opened to dispose of the projected volumes of captured $CO_2$. Four scenarios defined by two factors are specified. The two factors are activity in EOR and availability of captured $CO_2$. Under all scenarios, EOR operations can accommodate captured $CO_2$ through approximately 2025, after which significant development of geologic storage sites is needed.

## Purpose of Scenario Analysis

The development scenarios for CCS are intended to illustrate the potential bounds on supply and demand of $CO_2$ and shared services and equipment with the oil and gas sector. They are not intended to predict the future. By establishing a range of potential futures, we are able to indicate the conditions under which there are constraints regarding the development of geologic storage due to increased demand for auxiliary services.

## Two Principal Factors

Two principal factors define scenarios affecting the development of CCS. The first is the increased availability of anthropogenic $CO_2$ for EOR and geologic storage. To support existing and expanded EOR operations, several $CO_2$ capture projects are either currently operating or in planning and construction. These projects will deliver a relatively small amount of $CO_2$ in comparison to $CO_2$ made available from natural sources. Increased supplies of anthropogenic $CO_2$ would result from a requirement to reduce emissions of $CO_2$ and would drive the expansion of the $CO_2$ pipeline infrastructure to bring captured $CO_2$ to EOR fields or geologic storage sites (Dooley, Dahowski, and Davidson, 2009).

The second factor is the pace of activity in oil and gas exploration and development. Market prices are a key driver of this activity: high prices spur additional development, especially EOR operations, which become economical as prices of petroleum increase. Technological advances can help drive development: Advances in directional drilling and hydraulic fracturing have helped spur the development of the Marcellus Shale in Pennsylvania, West Virginia, and eastern Ohio; the Bakken Shale in North Dakota; and the Barnett and Haynesville-Bossier shales in Texas and Louisiana. As a result of a higher pace of activity in oil and gas development, the availability of oil and gas support services declines in the near term and the prices of those services increase. However, the market adjusts over time, as additional drilling equipment and services become available in response to increased demand.

## CCS Development Scenarios

By taking the two principal factors affecting development of CCS, we define four scenarios. Table 3.1 presents the scenarios. When oil and gas development proceeds slowly and anthropogenic supplies of $CO_2$ are limited, we expect EOR operations to decline. In this scenario, geologic storage activities would face little competition for resources, but at the same time such a scenario would correspond to a situation where there is no national requirement to reduce $CO_2$ emissions. Current conditions are characterized by limited supplies of anthropogenic $CO_2$ and a fast pace of oil and gas development (Scenario 2). Currently, the limited availability of $CO_2$ from natural deposits inhibits the expansion of EOR operations.[1]

In the scenarios described in the bottom row of Table 3.1, anthropogenic supplies of $CO_2$ are readily available, possibly as a result of a national requirement to reduce $CO_2$ emissions. If oil and gas development is slow, the shared equipment and services required to develop geologic storage sites will be readily available. Alternatively, if the pace of oil and gas development is high, there may be sufficient $CO_2$ for both geologic storage and EOR operations, but geologic storage applications will have to compete for equipment and services with the oil and gas industry.

### Potential Availability of Captured $CO_2$

We first estimate the amount of $CO_2$ that could be captured and made available for transportation to EOR or geologic storage sites. ICF, in a study for the Interstate Natural Gas Association of America, compared alternative policies that would lead to the deployment of systems to capture $CO_2$ (ICF International, 2009). Based on the estimated effects of these policies, ICF estimated low and high cases for the availability of captured $CO_2$. We plot the two cases in Figure 3.1; we have modified ICF's estimates, shifting estimates of available $CO_2$ by three years for the 2012 estimate and by five years for the estimates for 2015 and later. We shifted

**Table 3.1**
**Scenarios Affecting the Development of $CO_2$**

| Supply Conditions | Slow Pace of Oil and Gas Development | Fast Pace of Oil and Gas Development |
|---|---|---|
| Limited anthropogenic supplies of $CO_2$ are available | Scenario 1: Declining activity in EOR | Scenario 2: EOR activity at current or slightly increased levels, limited by natural sources of $CO_2$; little motivation to use EOR for storage; geologic storage develops slowly if at all |
| Significant supplies of anthropogenic $CO_2$ are available | Scenario 3: Unimpeded development of geologic storage | Scenario 4: Sufficient supplies of $CO_2$ for EOR and geologic storage; competition for shared services |

Source: RAND analysis.

[1] Interview with Michael L. Godec, Vice President, Advanced Resources International, Arlington, Va., February 21, 2012.

**Figure 3.1**
**Potential Supplies of $CO_2$ Under Two Cases**

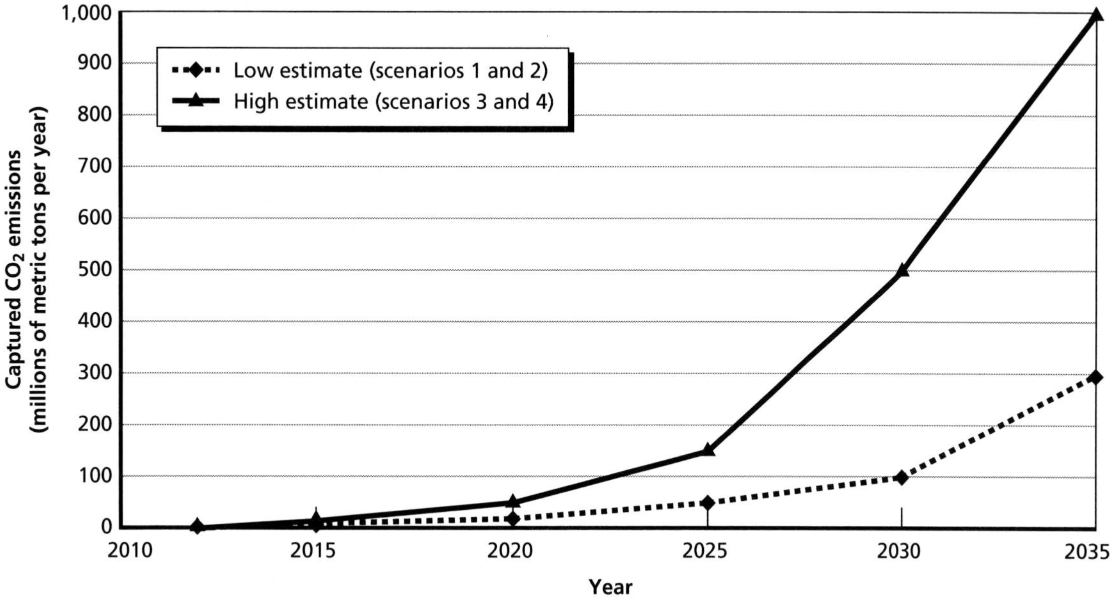

SOURCE: ICF International, 2009.
RAND *TR1300-3.1*

the policy implementation dates because of continued delays in enacting legislation or policy requiring the reduction of emissions of $CO_2$ from coal-fired power plants.[2] We have added some potential supplies of anthropogenic $CO_2$ for the years 2015, 2020, and 2025 to represent the potential availability of additional supplies of $CO_2$ from gas processing in which $CO_2$ has already been separated. Sixteen million metric tons of $CO_2$ are already captured from industrial facilities in the United States annually (NETL, 2011a).

According to the ICF estimates, the quantity of $CO_2$ available from capture systems in 2035 might range from 300 million to 1 billion metric tons annually, equivalent to captured emissions from 150 to 300 gigawatts of coal-fired power, depending on the efficiency of capture and a range of other factors.

In the remainder of the analysis, we will use the low estimate to represent available $CO_2$ in Scenarios 1 and 2, and the high estimate to represent available $CO_2$ in Scenarios 3 and 4.

### Potential Demand for $CO_2$ for EOR Operations

For EOR operations, a 2011 report by NETL estimated the potential effects of a set of similar scenarios (NETL, 2011a). The primary data source for the study is a nationwide database of oil reservoirs comprising approximately three-quarters of U.S. reserves. The potential production of each field through $CO_2$–EOR depends on the price of petroleum and the cost of $CO_2$. These two variables are analogous to the variables by which we specify scenarios in Table 3.1. As discussed above, the price of petroleum and the pace of activity of oil and gas development are correlated. The cost of $CO_2$ is a surrogate for its availability. The bounding scenarios considered in that study are listed in Table 3.2.

---

[2] Also, we would also expect that, where available and economical, fuel switching from coal to natural gas would constitute a near-term compliance strategy.

**Table 3.2**
**Oil and $CO_2$ Price Scenarios**

| Price of Petroleum | $60/barrel | $110/barrel |
|---|---|---|
| High price of $CO_2$, $/thousand cubic feet ($/metric ton) | $1.8 ($34) | $3.3 ($62) |
| Low price of $CO_2$, $/thousand cubic feet ($/metric ton) | $1.2 ($23) | $2.2 ($42) |

SOURCE: NETL, 2011a.

Based on the oil and $CO_2$ price scenarios, the study then estimated the potential demand for $CO_2$ to support EOR operations. If $CO_2$ prices fall below those assumed by the study, the demand for $CO_2$ would rise. In particular, EOR operators might apply advanced methods of $CO_2$ flooding in which higher volumes of $CO_2$ are injected into the reservoir to recover additional oil. These are the assumptions applied by NETL in their analysis, so the potential demand for $CO_2$ is higher than if existing methods for $CO_2$–EOR were employed. The results of these analyses with respect to $CO_2$ demand appear in Table 3.3. Table 3.4 illustrates the parameters used to estimate $CO_2$ demand.

NETL's estimates (2011a) assume the widespread application of advanced EOR techniques that employ significantly more $CO_2$ than would be used today. Since these techniques have not yet been fully developed, we adjust downward the potential demand for $CO_2$ by EOR operators assuming deployment of such techniques. In particular, for scenarios in which prices of $CO_2$ are high (supplies are limited), we adjust demand downward by 50 percent. For scenarios where the price of $CO_2$ is low (supplies are plentiful), we adjust demand downward by 33 percent.[3]

NETL (2011a) assumes that $CO_2$ would be purchased for use in EOR operations over a 50-year time period. This assumption corresponds to Scenarios 2 and 4 of this study, in which the pace of oil development is high. In Scenarios 1 and 3, where the pace of oil development is lower, we assume less-intense activity and a 75-year period over which EOR operations inject $CO_2$. To estimate demand for $CO_2$ over time, we assume that growth in demand for $CO_2$ by EOR operators grows linearly over the assumed time periods, which is consistent with past trends.

**Table 3.3**
**Estimated $CO_2$ Demand for EOR Operations (in Billions of Metric Tons)**

| Price of Petroleum | $60/barrel | $110/barrel |
|---|---|---|
| Demand for $CO_2$ in high $CO_2$ price case | 16 | 21 |
| Demand for $CO_2$ in low $CO_2$ price case | 18 | 22 |

SOURCE: NETL, 2011a.

[3] NETL (2011a) assumes that "next generation" $CO_2$–EOR methods inject 1.5 times the hydrocarbon pore volume. "State of the art" $CO_2$–EOR today injects $CO_2$ quantities approximately equal to the hydrocarbon pore volume, or one-third less $CO_2$ than would be injected in the "next generation" case. Many current $CO_2$–EOR operations inject $CO_2$ at approximately 0.8 times the hydrocarbon pore volume (NETL, 2011a), which is approximately one-half of the $CO_2$ that would be injected in the "next generation" case.

**Table 3.4**
**Parameters Used to Estimate Demand for $CO_2$ by EOR Operations**

| Scenario from NETL (2011a) | Scenario 1 (Low price of oil, high price of $CO_2$) | Scenario 2 (High price of oil, high price of $CO_2$) | Scenario 3 (Low price of oil, low price of $CO_2$) | Scenario 4 (High price of oil, low price of $CO_2$) |
|---|---|---|---|---|
| Estimate of cumulative $CO_2$ demand for EOR (billion metric tons) | 16 | 21 | 18 | 22 |
| Adjusted cumulative demand for $CO_2$ (billion metric tons) | 8 | 11 | 12 | 15 |
| Years of $CO_2$ purchases for EOR operations | 75 | 50 | 75 | 50 |

SOURCE: NETL, 2011a.

CHAPTER FOUR

# The Capacity of the $CO_2$ Storage Industrial Base to Respond to the Development Scenarios

In this chapter, we quantify the capacity of the $CO_2$ storage industrial base to respond to the pressures posed by the scenarios developed in Chapter Three. We evaluate potential constraints on pipeline construction, EOR development, and the deployment of geologic storage. The response of the industrial base is compared according to historical activities of the sector, as well as according to equipment and labor needs.

## Infrastructure for Transporting Captured $CO_2$

In its analysis of infrastructure needs to support $CO_2$ storage, ICF estimated the capacity and extent of pipelines to transport captured $CO_2$ to potential storage sites (ICF International, 2009). ICF International describes four scenarios: two for low and high rates of CCS (using the low and high supplies of $CO_2$), and two for lesser and greater applications of $CO_2$ for EOR operations. The pipeline requirements derived by ICF are shown in Figure 4.1, which presents total mileage of $CO_2$ pipelines for ICF's four scenarios. We have added the 4,500 miles of existing $CO_2$ pipelines to the ICF estimates. We assign these estimates from ICF to Scenarios 1 through 4, as specified in Chapter Three.

In the low CCS scenarios (Scenarios 1 and 2), the total mileage of $CO_2$ pipelines grows from 4,500 miles today to 9,500–12,000 miles in 2035, or an annualized rate of addition of about 200 to 350 miles per year. In the high CCS scenarios, the total mileage of $CO_2$ pipelines grows to 24,000 in the scenario in which there is little additional EOR activity (Scenario 3) and 40,000 in the scenario where there is significant additional EOR activity (Scenario 4), representing an annualized addition rate of about 850 to 1,500 miles per year, respectively. These annualized additions assume this rate of construction occurs even in the early years of deployment. A more conservative bounding assumption would be that annualized deployment occurs over the ten-year high-deployment period of 2025–2035. Under these conditions, annualized $CO_2$ pipeline construction would need to be about 1,700 to 3,200 miles per year to meet Scenario 3 and Scenario 4, respectively. This illustrates the impact of compressed timing or rapid growth on annual pipeline construction requirements.

Other estimates of the pipeline requirements to transport captured $CO_2$ have been published. For example, Dooley, Dahowski, and Davidson (2009) estimate pipeline requirements under two scenarios related to stabilizing atmospheric concentrations of $CO_2$ by 2050: In 2030, they estimate that between approximately 10,000 and 22,000 miles of $CO_2$ pipelines

**Figure 4.1**
**Pipeline Transportation Infrastructure Under Four Scenarios**

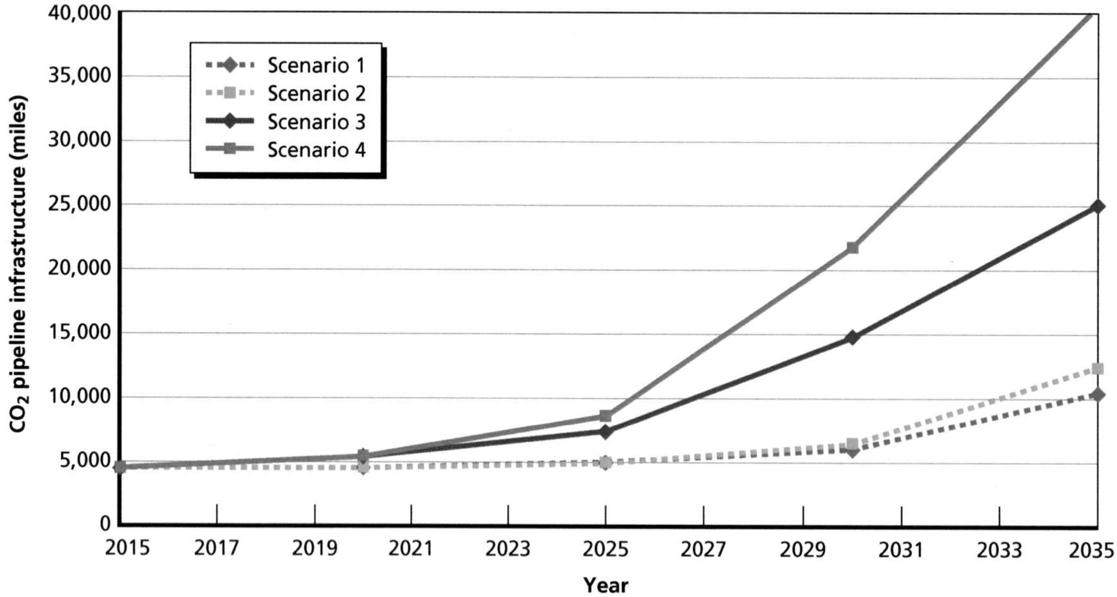


SOURCE: ICF International, 2009.
RAND *TR1300-4.1*

are required, growing to between 16,000 and 28,000 miles in 2050. These cases fall between the scenarios illustrated in Figure 4.1.

The required additions of $CO_2$ pipelines can be compared with historical $CO_2$ and natural gas pipeline additions for perspective on how the industrial base has responded previously. Existing $CO_2$ pipelines were largely constructed to connect a single or few sources of $CO_2$ to a single or few EOR users. A transmission network was neither planned nor established. The largest recent annual addition of $CO_2$ pipelines was 625 miles in 2005. If this rate were replicated, the $CO_2$ pipeline demands of the low carbon-storage scenario and the high scenarios where there is little EOR activity (Scenario 3) would likely be met. However, the high scenario where there is extensive EOR activity (Scenario 4) could require an annualized construction rate of about 2.5 to 5 times the length of $CO_2$ pipelines constructed in 2005, depending on the starting point of major $CO_2$ pipeline deployment.

Recent annual additions to natural gas transmission pipelines are more than double the required rate of $CO_2$ pipeline additions in Scenario 4, which requires the most additional pipeline infrastructure. As shown in Figure 4.2, the highest recent rate of annual natural gas pipeline additions was more than 3,500 miles in 2008 and 2009. Moreover, in the 1970s, the growth of natural gas transmission pipelines was double that of later decades (Dooley, Dahowski, and Davidson, 2009). While there are some differences between $CO_2$ pipelines and natural gas pipelines, these different types of pipelines largely share the same industrial base. Therefore, the required scale of additions to the $CO_2$ pipeline is unlikely to be an impediment to the development of $CO_2$ storage: The industrial base has shown the ability to develop pipelines at greater rates in the past, and the need for pipelines would not be sudden, but would grow as capture systems are deployed, providing time for the industrial base to respond. However, as illustrated, in the event rapid deployment occurred from 2025–2035, annualized $CO_2$

**Figure 4.2**
**U.S. Natural Gas Pipeline Installation History, 1997–2011**

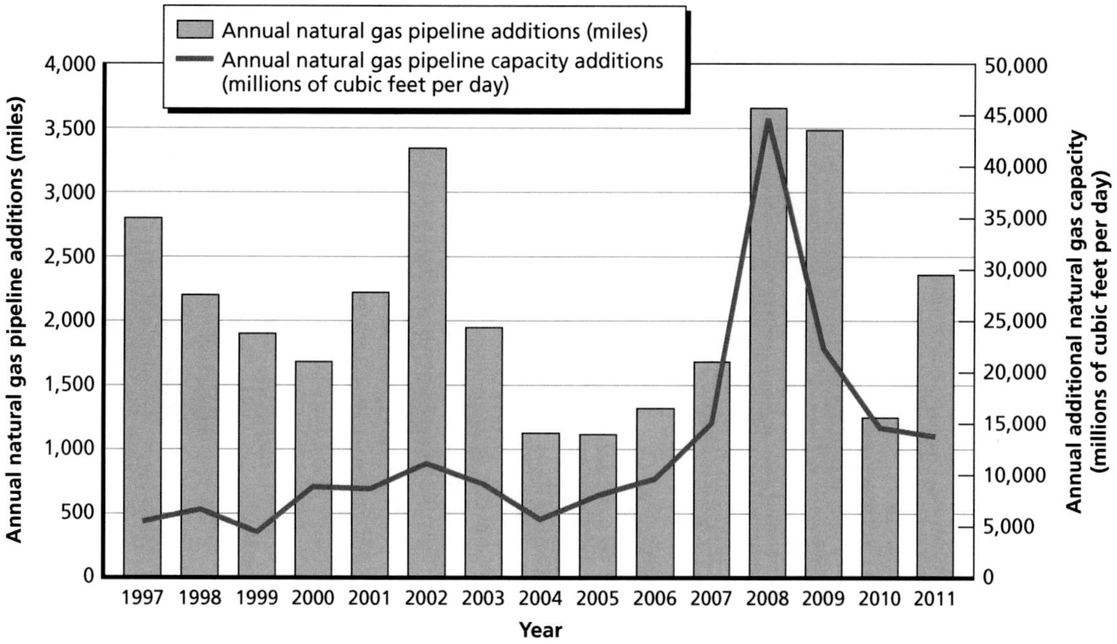

SOURCES: EIA, undated; PHMSA, 2012a; personal communication with Tu T. Tran, energy economist, EIA, 2012.
RAND *TR1300-4.2*

pipeline deployment would be approximately the same as recent natural gas pipeline additions. We note that the starting points of future $CO_2$ pipelines will be at power plants, rather than the mostly rural natural gas wells. Differences in geographic and population characteristics between power plants and natural gas wells could affect the timing of permitting and construction of $CO_2$ pipelines.

## Disposition of Captured $CO_2$ for EOR and Geologic Storage

Because EOR operations are mature, we assume that they will be the first to use available $CO_2$. NETL (2011a) estimated several cases for the potential for EOR to store $CO_2$. In contrast to the analysis by ICF (2009), NETL based their analysis on scenarios of future prices of $CO_2$ and petroleum. We assume that low $CO_2$ prices correspond to the high availability of $CO_2$. We assume that the price of petroleum drives activity in EOR: lower oil prices reduce activity in EOR while higher oil prices increase it. Next, we need to estimate the amount of $CO_2$ that would be used to support EOR operations. Here we make a simplifying assumption: because EOR technologies are mature and because there is a financial incentive to produce additional petroleum, we assume that demand is satisfied for EOR operations prior to $CO_2$ being diverted to geologic storage. Depending on the availability of transportation infrastructure and local geologic conditions, certain regions of the country may see early geologic storage activity, rather than expanded EOR operations. For example, in the Appalachian region, there are relatively few opportunities for EOR by $CO_2$ flooding; NETL (2011a) estimates that there are 1.3 billion barrels of additional petroleum that could be produced by EOR in this region, which contrasts with

14.6 billion barrels of economically recoverable petroleum in the Permian basin in west Texas and New Mexico.

The implications of this assumption on our estimate of the demands on the industrial base for carbon storage are small. In general, this assumption will overestimate the amount of $CO_2$ that is used for EOR operations. Also, by implication, we will overestimate the number of EOR sites in relation to the number of geologic storage sites. As will be discussed later, the surface infrastructure requirements for EOR and geologic storage are expected to be similar, so our estimate of the overall demand for services—well drilling, seismic, and $CO_2$ distribution pipelines, for example—will also be similar.

In Figure 4.3, we depict demand for $CO_2$ for EOR and geologic storage over time for the four analytical scenarios. Because there is relatively little incremental $CO_2$ available in the near term from $CO_2$ capture, commercial geologic storage does not begin to displace EOR operations until after 2020 in the scenarios where there is relatively low EOR activity (Scenarios 1 and 3). In scenarios where there is increased EOR activity (Scenarios 2 and 4), requirements for geologic storage are minimal until after 2025 or 2030.

**Figure 4.3**
**Disposition of Captured $CO_2$ Under Four Scenarios**

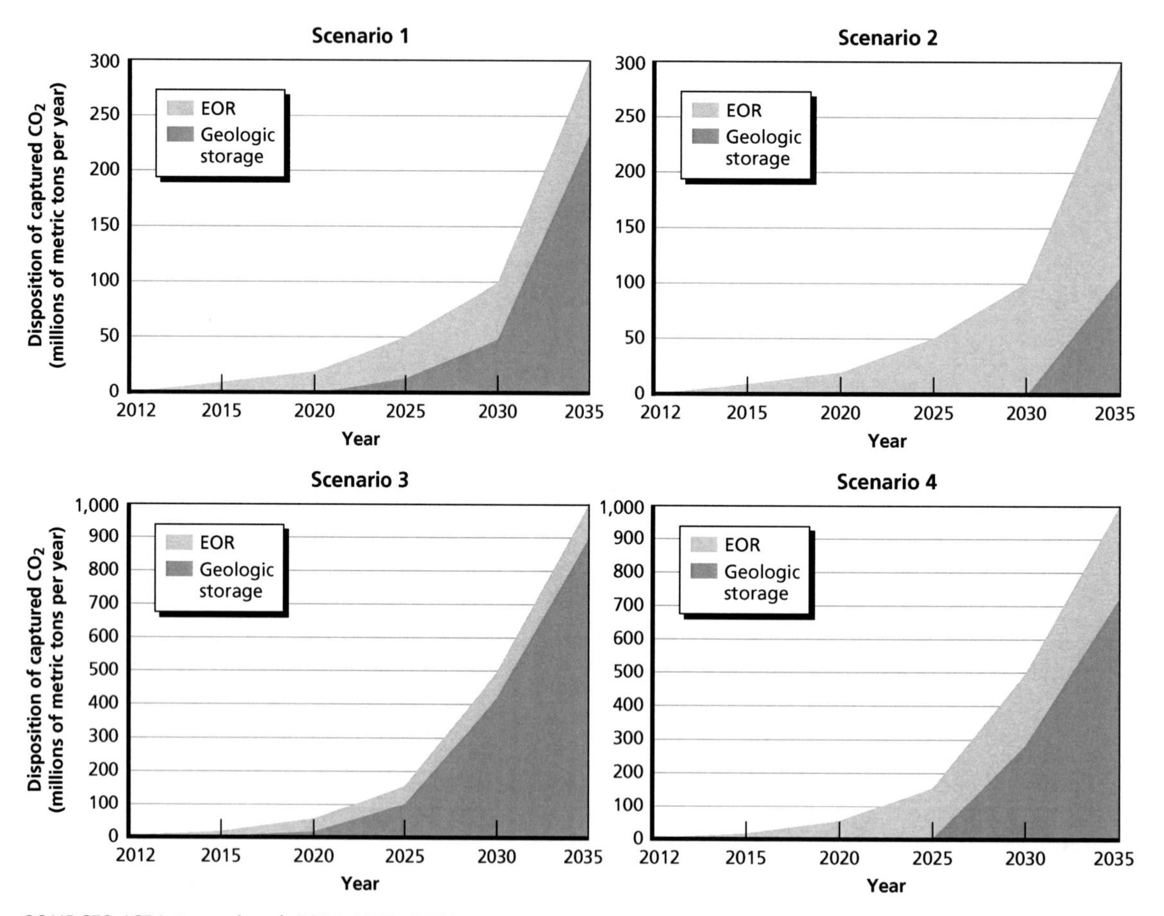


SOURCES: ICF International, 2009; NETL, 2011a.
RAND *TR1300-4.3*

## Industrial Base Requirements for $CO_2$ Storage

The industrial base requirements to support $CO_2$ storage by either EOR operations or geologic storage activities are proportional to the number of new EOR flooding projects or commercial-scale geologic storage sites. There is a high degree of variation in the scale of EOR operations. Some fields are relatively small and have a handful of injecting and producing wells; other fields are relatively large and have hundreds of injecting and producing wells (Advanced Resources International, 2006). Figure 4.4 depicts the production of petroleum attributed to enhanced techniques by the number of total wells at the EOR site.

Figure 4.4 illustrates the wide range in the scale of operations at active $CO_2$–EOR sites. The total number of injecting and producing wells at active sites ranges from two to more than 1,600. Incremental production of petroleum ranges from a minimum of ten barrels per day to more than 26,000. The relevant activities with respect to the carbon storage industrial base are related to the number of wells at the site. While each EOR site is unique, a portion of the existing wells at a typical site are repurposed into either injecting or producing wells, another portion of the existing wells are plugged, and some new wells are drilled. The development of wells and related infrastructure for injecting $CO_2$ and recovering produced oil occurs over many years. Among active $CO_2$–EOR sites today, the average number of total wells onsite is 153, but that figure is skewed due to a relatively small number of very large sites (Figure 4.5). The median number of wells at a site is 58.

**Figure 4.4**
**Enhanced Oil Production Versus Number of Injecting and Producing Wells at Onshore U.S. $CO_2$–EOR Sites**

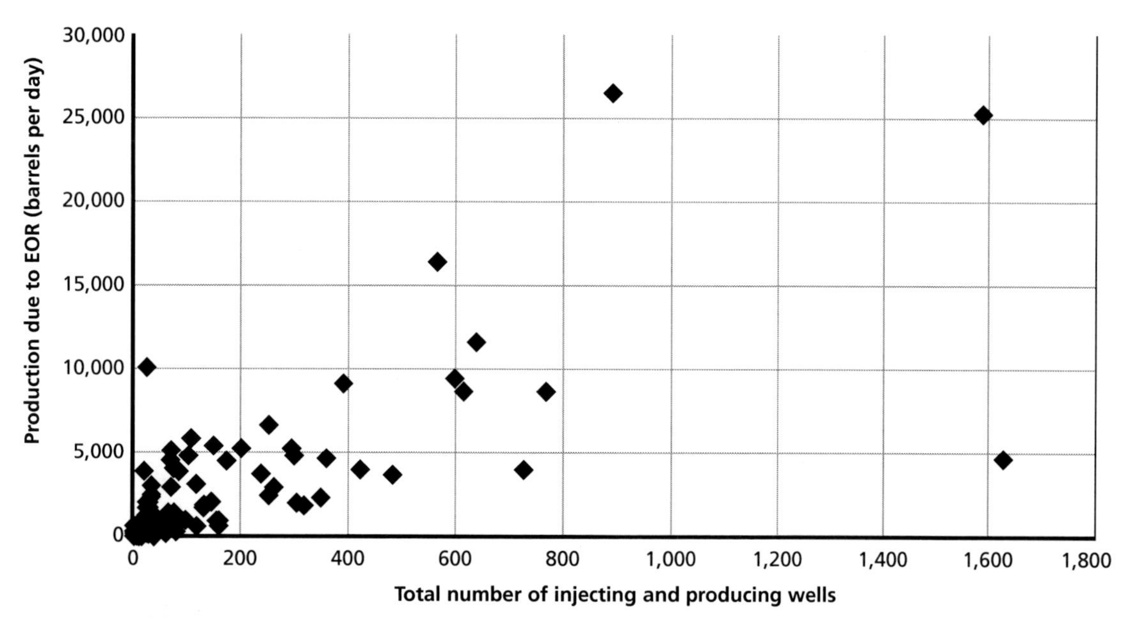


SOURCE: Moritis, 2010.
RAND *TR1300-4.4*

**Figure 4.5**
**Histogram of Number of Injecting and Producing Wells at an EOR Site**

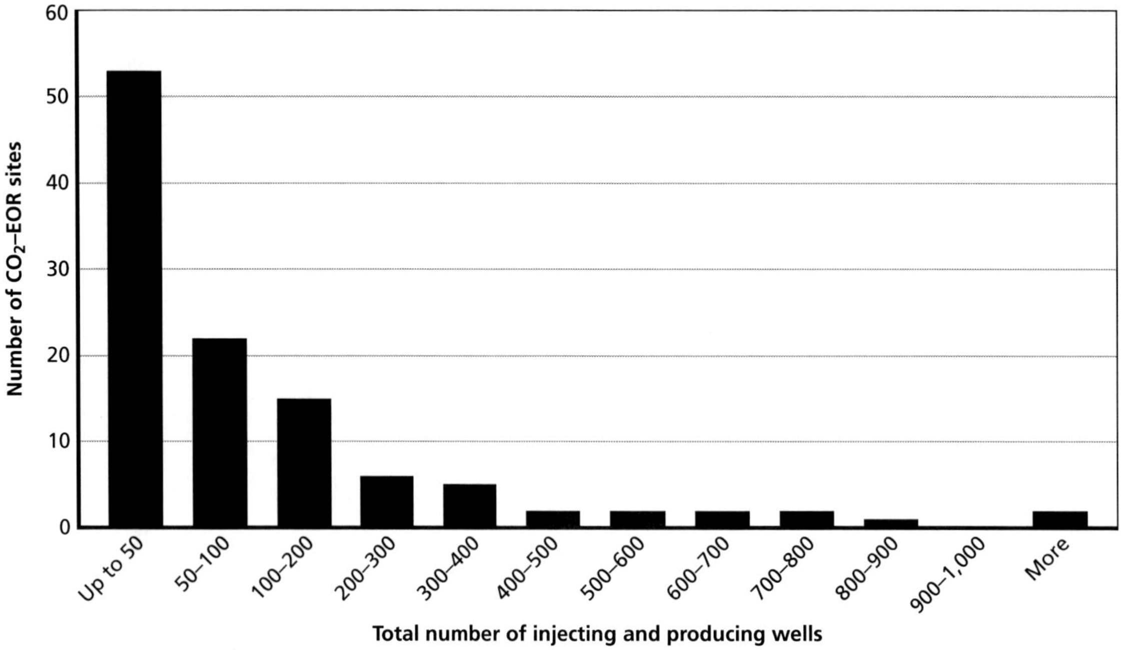

Number of $CO_2$–EOR sites

Total number of injecting and producing wells

SOURCE: Moritis, 2010.
RAND *TR1300-4.5*

We use EOR operations in the Permian basin as a benchmark for the scale of $CO_2$ stored in EOR operations.[1] This assumption is made for two reasons. First, this is because EOR operations are most developed in the Permian basin: 30 of the 114 active $CO_2$–EOR projects reported in 2011 are in the Permian basin (Moritis, 2010). Second, more data are publicly available regarding specific EOR activities in this region. It is important to note that although reservoirs within a geographic basin may have similar properties, the redevelopment of fields using EOR depends critically on how the field was developed initially and the scale and scope of other enhanced recovery operations that occurred on site. When we extrapolate this assumption to development nationwide, the implication, on average, is that the amount of $CO_2$ used and stored in an EOR operation as a function of the number of wells that need to be drilled, redeveloped, and plugged is the same. We will address the sensitivity of our results to this assumption.

Figure 4.6 illustrates the assumed injection profiles of typical geologic storage sites and $CO_2$ purchases for EOR projects. We assume that the injection rate at geologic storage sites is 2 million metric tons of $CO_2$ per year for 20 years. In its cost model for geologic storage, NETL (2012e) estimates that a storage site with two injection wells could inject 4.1 million metric tons per year for 30 years. The implication of our assumption is that injection wells would accept approximately one-half the $CO_2$ as modeled by NETL and would operate for less time. Therefore, our results should show approximately twice the number of $CO_2$ sites as

[1] While we focus on the injection of $CO_2$ into the formation to enhance the recovery of petroleum, it is important to note that in a typical $CO_2$–EOR operation, water is injected periodically in a process known as "water alternating with gas" to help control the movement of the $CO_2$ in the formation.

**Figure 4.6**
**Annual and Cumulative Stored $CO_2$ for EOR and Geologic Storage**

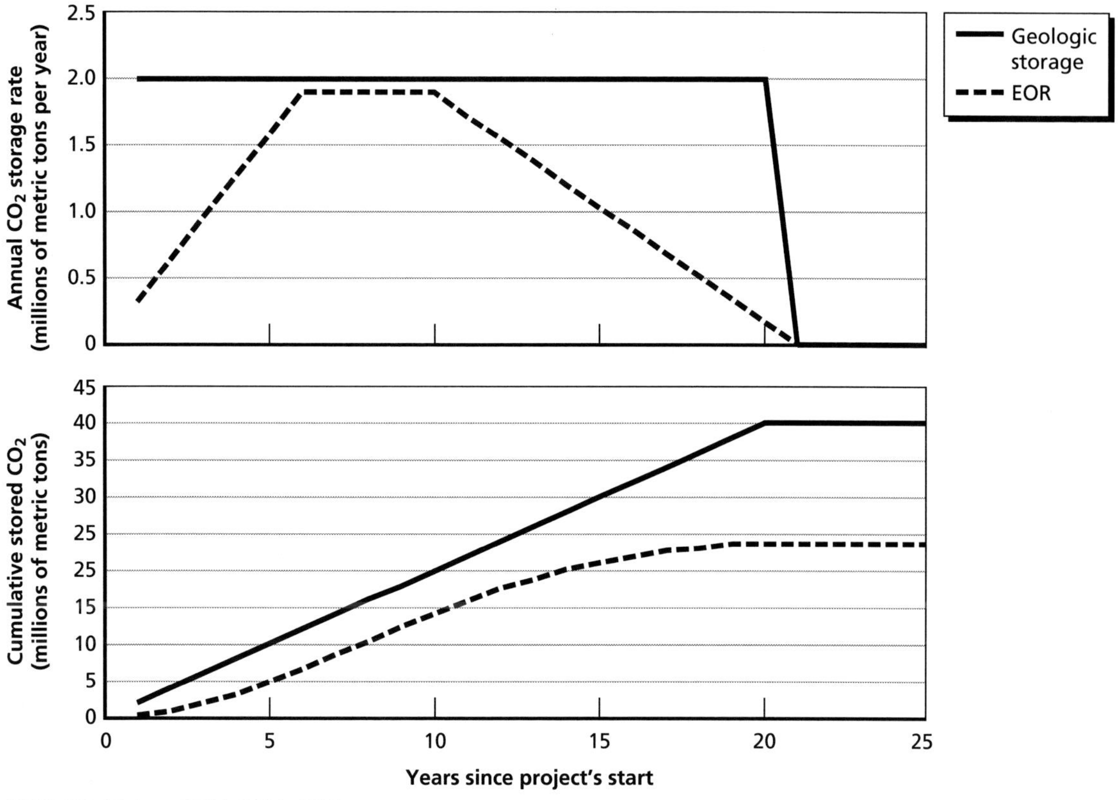


SOURCES: Melzer, 2011; NETL, 2012e.
RAND *TR1300-4.6*

would be the case under NETL's assumptions. The total amount of $CO_2$ stored at the site is 40 million metric tons.

For EOR projects, we assume the profile of $CO_2$ purchases for a typical large EOR site in the Permian basin (Melzer, 2011). As shown in Figure 4.6, there is a five-year ramp-up period over which the rate of injection of purchased $CO_2$ rises to 1.9 million tons per year. $CO_2$ purchases remain at this level for the next five years, and decline to zero over the next ten years. The EOR project will continue to reinject $CO_2$ that it separates from produced oil, but will not purchase any new $CO_2$. We assume that the project ultimately stores all purchased $CO_2$, approximately 24 million metric tons over 20 years. This is the general profile of an existing EOR operation. NETL (2011a) provides an overview of "next generation" EOR approaches, which may use more $CO_2$ per unit of produced hydrocarbon. If application of advanced $CO_2$–EOR methods becomes prevalent in the future, then we may overestimate the number of EOR projects required to store a given volume of $CO_2$.

The final step in our analysis is to estimate the number of active EOR and geologic storage projects. Beginning in 2015, we first determine the number of EOR projects needed to dispose of the captured $CO_2$, as illustrated in Figure 4.3 for each scenario. We then do the same for the $CO_2$ allocated to geologic storage. Both estimates are rounded up to the nearest integer. To estimate the number of projects that need to be started in 2020, we estimate the amount of $CO_2$ that existing projects could dispose of. Then we determine the number of new projects that would be needed to dispose of any remaining $CO_2$. We repeat this recursive procedure for the remaining years. The results are depicted in Figure 4.7. Numerical values are listed in Table 4.1.

Based on the given number of projects, we estimate the requirements that such development would place on the $CO_2$ storage industrial base. As noted in the discussion above, the estimate of the number of active $CO_2$–EOR and geologic storage sites that would be needed to store captured $CO_2$ is uncertain. The estimates provide a basis on which we can extrapolate the requirements of the industrial base. The principal concern regarding implications for the carbon storage industrial base would be that we systematically over- or underestimated the scale of $CO_2$–EOR or geologic storage sites in terms of infrastructure requirements at the surface or the amount of $CO_2$ stored. The potential implications involve the following:

- **Pipelines.** If the typical scale of a $CO_2$–EOR or geologic storage operation were smaller in terms of the total amount of $CO_2$ stored, then there would be an increase in the need for pipelines to transport $CO_2$ to individual sites. The increase in pipeline mileage would

**Figure 4.7**
**Number of Active Large EOR or Geologic Storage Projects Required to Store Captured $CO_2$**

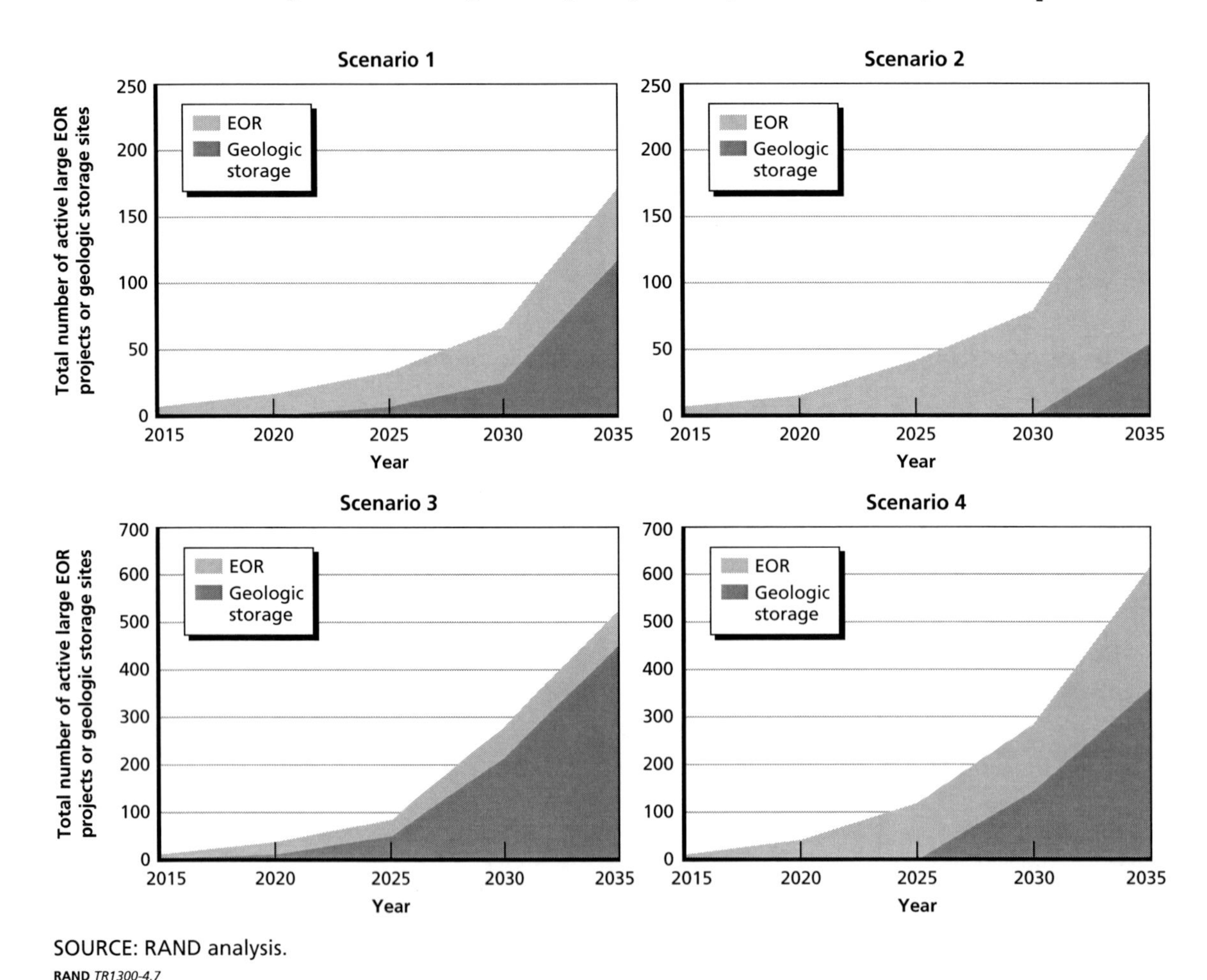


SOURCE: RAND analysis.
RAND *TR1300-4.7*

**Table 4.1**
**Estimated Number of Active Large EOR or Geologic Storage Projects Required to Store Captured $CO_2$**

| | Scenario 1 Low Price of Oil, High Price of $CO_2$ | | Scenario 2 High Price of Oil, High Price of $CO_2$ | | Scenario 3 Low Price of Oil, Low Price of $CO_2$ | | Scenario 4 High Price of Oil, Low Price of $CO_2$ | |
|---|---|---|---|---|---|---|---|---|
| Year | Geologic Storage | EOR | Geologic Storage | EOR | Geologic Storage | EOR | Geologic Storage | EOR |
| 2015 | 1 | 6 | 0 | 7 | 1 | 9 | 0 | 10 |
| 2020 | 1 | 16 | 0 | 15 | 9 | 27 | 0 | 43 |
| 2025 | 7 | 27 | 0 | 42 | 48 | 36 | 0 | 120 |
| 2030 | 25 | 42 | 0 | 79 | 210 | 68 | 140 | 140 |
| 2035 | 120 | 54 | 54 | 160 | 450 | 75 | 360 | 250 |

NOTE: RAND analysis; scenarios are derived from NETL, 2011a.

occur at the edges of the transmission network, as additional pipelines would have to be built to service additional sites in the same geologic formations.

- **Drilling rigs and field services.** The number of drilling rigs that would be required is directly proportional to the number of wells. The more critical parameter from the perspective of the carbon storage industrial base is the number of wells required per unit of $CO_2$ stored, which we overestimate. With respect to geologic storage, we assume that twice the number of wells are required per unit of $CO_2$ stored than is assumed by NETL (2012e). With respect to $CO_2$–EOR sites, our estimates are based on current operations, which do not take into account the potential development of operations that employ increased volumes of $CO_2$. Since many of the other services at a site are a function of the number of wells to be maintained, similar observations apply.

## Discussion

### *EOR Operations Can Store Available $CO_2$ Until Approximately 2025*

In all four scenarios, between six and ten new large EOR projects are required to dispose of captured $CO_2$ in 2015. From 2012 to 2015, this corresponds to two to three large projects per year, which is completely within the capabilities of the industry. In 2020, the number of new EOR projects ranges from eight to 33, or two to seven per year; over the same 2015–2020 time period, geologic storage activity is minimal. Between 2020 and 2025, the number of additional geologic storage projects opened in Scenario 3 is approximately eight per year, whereas additional EOR sites ranges from one to 14 per year. After 2025, captured $CO_2$ supplies increase significantly, requiring more activity in both EOR and geologic storage.

### *Growth Requirements for EOR Are Largely Within Historical Experience*

As discussed in Chapter Three, EOR activity has increased significantly in the United States in recent years, averaging seven new projects per year since 2006. For Scenarios 1 and 3 (those with less EOR activity), the maximum number of projects added per five-year period is 32 (Scenario 3 between 2025 and 2030). The maximum number of new EOR projects in a five-year period for Scenarios 2 and 4 (those with more EOR activity) is 120, which occurs between 2030 and 2035 in Scenario 4. This is an average of 24 projects per year, three times the pace of recent development. Interviewees told us, however, that development in recent years has been constrained by the cost and lack of supplies of $CO_2$; if significant quantities of $CO_2$ were to become available, the pace of growth would likely be faster. As per NETL (2011a), next-generation EOR methods have the potential to use significantly more $CO_2$ than existing methods on which our model is based, so the number of projects may be overestimated if such methods come into widespread use.

#### *Significant Expansion of Geologic Storage Capacity Is Needed After 2025*

In most scenarios, the ability of EOR operations to accommodate most of the potentially available $CO_2$ diminishes after 2025. Between 2025 and 2030, 18 additional geologic storage sites are needed in Scenario 1 and 160 in Scenario 3. On average, between four and 32 new geologic sites would need to be added per year. Requirements for storage sites increase further in the 2030 to 2035 timeframe: The minimum is 54 sites for Scenario 2 and the maximum is 240 sites for Scenario 3, representing annual averages of 11 to 48 new sites per year. By contrast, the maximum number of EOR sites added per year is 24, which occurs in Scenario 4 between 2030 and 2035. This is more than triple the number of sites that have been developed in recent years, but the requirements that this would place on the industrial base are modest in comparison to the scale of oil and gas operations, as we will discuss.

#### *The Industrial Base Requirements to Support Storage of $CO_2$ Are Modest*

The equipment, labor, and services required to develop geologic storage and EOR sites to store the captured $CO_2$ represent relatively small increases in activity in comparison to the total supporting industry. Requirements to develop pipelines were discussed above. Here we discuss the number of rigs that would be required to drill and maintain wells, the number of seismic survey teams, and the skilled labor requirements to support expansion of the industry. In all cases, the incremental increases are relatively small and would not impede growth within the scale of activities illustrated in the four scenarios covered here. One potential barrier to further development of the industry that we have not considered in detail is the potential for delays in permitting new sites.

Number of wells and drilling rigs. Both geologic storage and EOR sites require a number of wells to be drilled or reworked. According to a cost model developed by NETL (2012e), a typical geologic storage site will have two injection wells and two in-formation monitoring wells. Additionally, eight shallower wells will be required to monitor the progression of the $CO_2$ plume and potential effects on groundwater supplies. We assume the following time table:

- Drilling and completing an injection well takes three months, with one month added to demobilize, move, and mobilize the rig.[2]
- Drilling and completing in-formation wells takes two months, again with one month added to move the rig.
- Drilling and completing above-seal monitoring and groundwater monitoring wells takes one month.
- Drilling and completing shallow monitoring wells takes two weeks.

---

[2] Rowe interview, 2011.

Thus we estimate approximately two total rig-years per site. In this analysis we do not make a distinction among sizes or weight capacity of rigs that might vary among sites.

For EOR, we base the rig requirements on the ongoing Hastings Fault Block A development in Mississippi (Davis et al., 2011). Reactivating this portion of the field requires 152 wells to be either drilled or to be worked over during a period of four years. Assuming that drilling and completing new wells requires one month and working over existing wells takes two weeks, we estimate the rig requirements to be approximately two rigs per year per site.

The industrial base has the capacity to support developing sufficient wells to support the development of carbon storage projects. Each geologic storage site requires approximately 14 wells to be drilled over the course of two years and each large EOR site will have to drill or work over approximately 40 wells per year. Recall that the highest rate of development of geologic storage sites is estimated to be 48 sites per year, which would require an average of 340 wells to be drilled per year. The highest rate of development of EOR sites is 24, which would require 960 wells to be drilled per year. Therefore, annual well development requirements to support carbon storage between 2030 and 2035 are approximately 1,300 per year. Between 2000 and 2011, the number of exploratory and developmental oil and gas wells drilled per year in the United States varied from 28,000 to 54,000 (EIA, 2012). Even if our estimates of well requirements are off by a factor of two, the requirements placed on the industrial base are less than a tenth of total demonstrated capacity.

We take the well requirements estimated above to derive the rig requirements for developing geologic storage and EOR sites for the four scenarios. These rig requirements appear in Figure 4.8. The number of rigs required across the scenarios in the 2030–2035 time period range from approximately 42 to 130.

Putting Figure 4.8 into context, consider Figure 4.9, which shows the number of well service and rotary drilling rigs in operation in the United States (EIA, 2012). Currently, there are almost 2,200 well-service rigs and almost 2,000 rotary rigs in operation. These figures have increased from 1,600 well-service rigs and 850 rotary rigs in operation in mid-2009. The maximum number of rigs that would be required to support development and geologic storage through 2035 is approximately 130—7 percent of the total number of active rotary rigs or 3 percent of the total number of rigs in operation. The current utilization rate of rigs in the United States is 67 percent, which is historically high for the industry, but indicates that additional capacity remains (Berkman and Stokes, 2011). As noted, our estimate for the number of rigs required is crude and could over- or underestimate the number needed to fulfill the needs for carbon storage. However, if our estimate were off by a factor of two, then 13 percent of rotary rigs or 6 percent of the total U.S. onshore capacity would need to be dedicated to carbon storage activities.

In our scenarios the projected pace of activity in geologic storage and EOR grows significantly after 2025, so comparisons of industrial base needs in 2035 to today's must account for the development of additional equipment. The oil and gas industry has been able to add new equipment as technology advances and demand rises: Since 2006, newly constructed rigs numbered between 131 and 349, indicating sufficient capacity to adjust to market demands, especially since carbon storage would be driven largely by long-term policy (Berkman and Stokes, 2011). Whether there is a shortage of rigs to support development of geologic storage

**Figure 4.8**
**Approximate Number of Active Drilling Rigs Required to Develop Geologic Storage and EOR Sites**

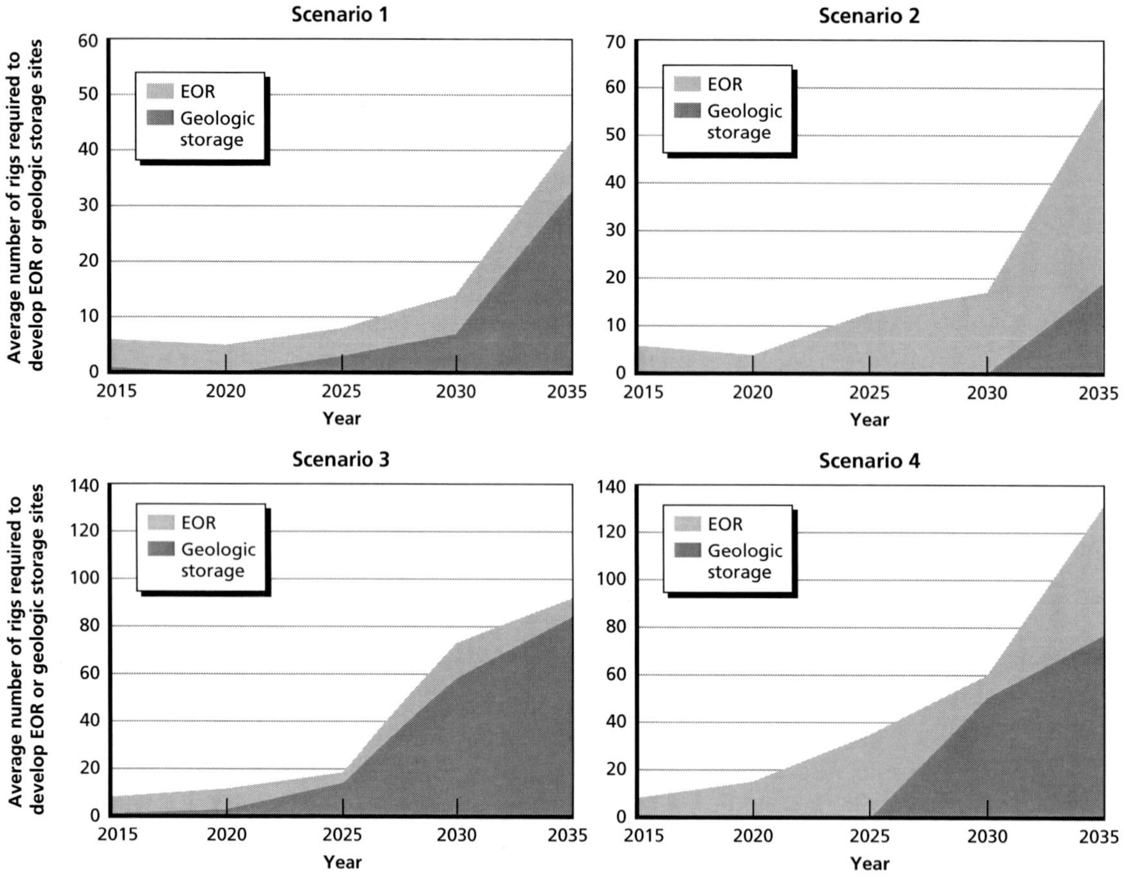

SOURCES: NETL, 2012e; Rowe interview, 2011.
RAND *TR1300-4.8*

or EOR sites in the future will depend on regional availability and rig capacity.[3] The current regional distribution of rigs is a result of demand and geology; most movements of rigs are intraregional, rather than interregional. In areas where geologic storage is to be developed and where there is currently little oil and gas exploration and development, shortages of appropriate rigs may occur during the early stages of development.

**Geologic surveying.** Geologic surveying is another key activity supporting the development of both geologic storage and EOR sites. One key component of a geologic survey is seismic imaging of the subsurface features. Seismic imaging is carried out by a crew using a truck to induce vibrations and sensors placed throughout the area to record the reflected waves. Depending on the accessibility of the area under review, seismic equipment may be delivered by helicopter. The data that are collected are then used to build a geologic model of the site. Assuming that the time on site for a geologic survey is one and one-half months per site, we can derive approximate requirements for the number of additional seismic crews needed to support development of additional geologic storage and EOR. Some geologic surveying occurs as part

[3] Rowe interview, 2011.

**Figure 4.9**
**Active Crude Oil or Natural Gas Drilling Rigs**

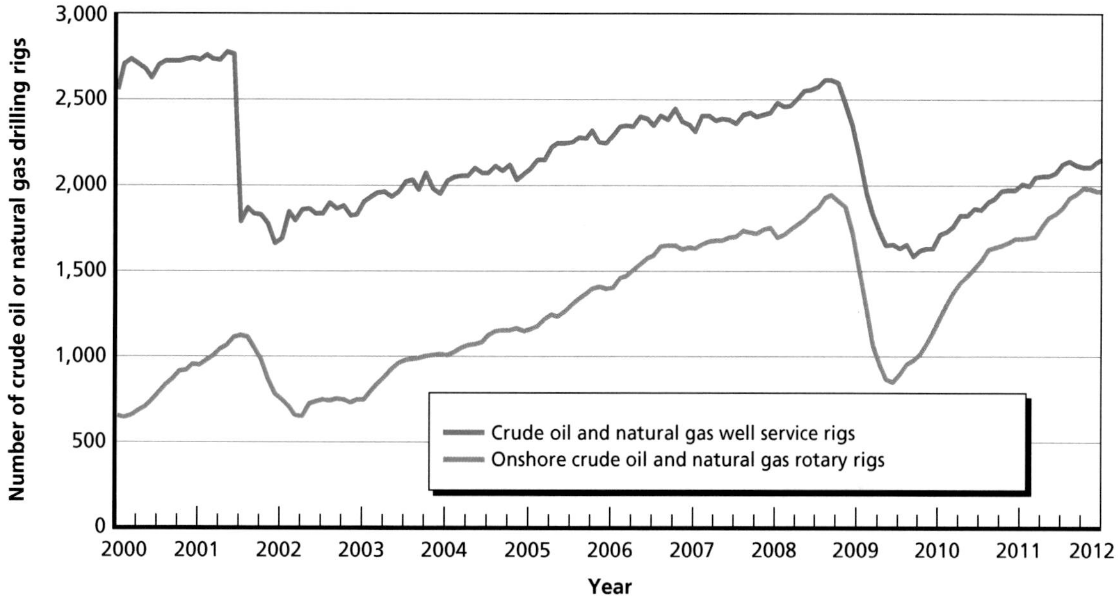

SOURCE: EIA, 2012.
RAND *TR1300-4.9*

of drilling operations through down-hole measurements;[4] we assume these services are carried out by the field services team.

The maximum number of additional seismic crews required to support development of geologic storage and EOR in 2035 ranges from four to 10 depending on the scenario (see Figure 4.10).

**Labor.** The additional labor needed to support the development of geologic storage sites is relatively modest. The cost model for geologic storage estimates that in the four years prior to opening a site, the average resource load is 25 percent for a geologist, 20 percent for an engineer, and 40 percent for a "landman" to negotiate and contract for the land rights for the site (American Association of Professional Landmen, 2012; NETL, 2012e). The American Association of Professional Landmen (2012) claims that there are approximately 13,000 people employed in the profession. We estimate the maximum number of geologic storage projects to be developed to be 240 over the five-year period from 2030 to 2035. Developing these projects would continuously employ approximately 50 geologists, 40 engineers, and 80 landmen. Continued operations at sites will employ additional people, but a small fraction of the total employed by the oil and gas and other supporting sectors.

[4] Rowe interview, 2011.

**Figure 4.10**
**Number of Active Seismic Surveying Teams Required to Support Development of Geologic Storage and EOR Under Four Scenarios**

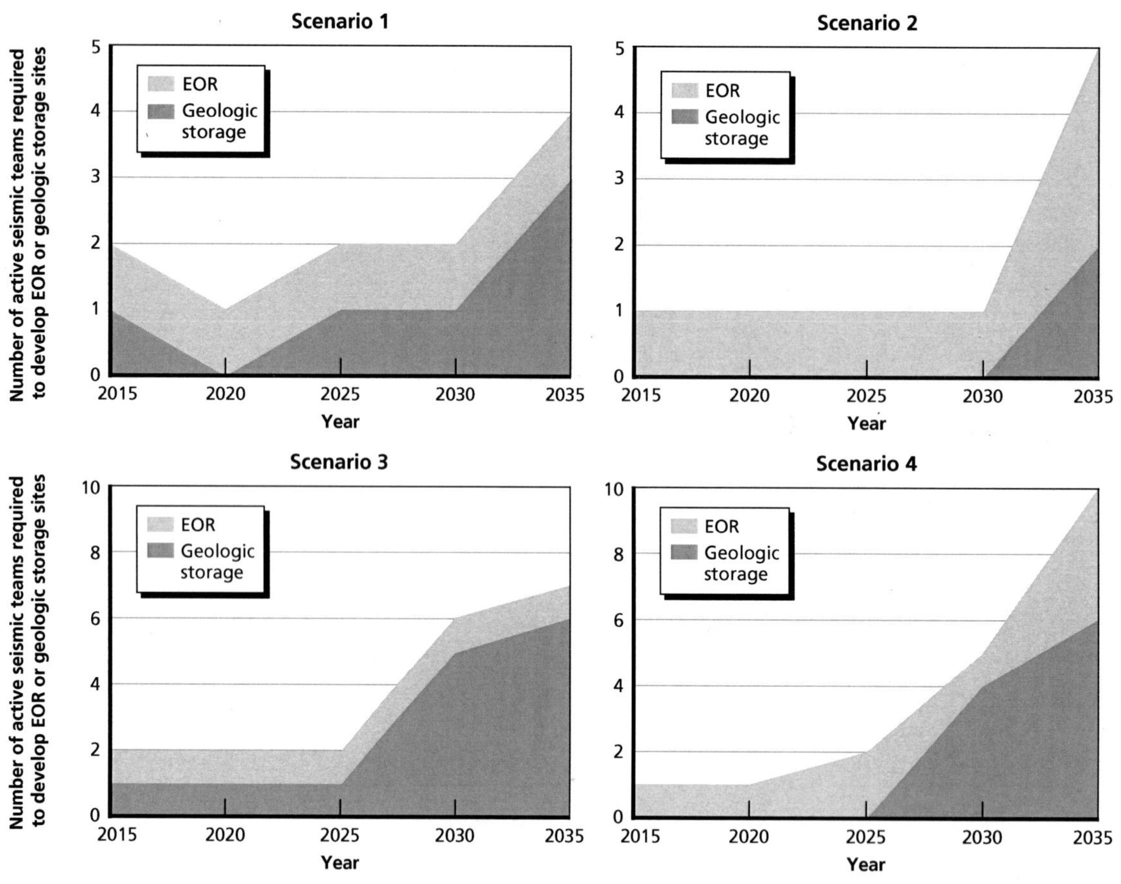


SOURCE: RAND analysis.
RAND *TR1300-4.10*

CHAPTER FIVE

# Findings and Implications

This chapter presents findings derived from our analysis of the $CO_2$ storage industrial base and the implications of those findings for the activities of the NETL CCS program. The $CO_2$ storage industrial base has a foundation based in the oil and gas industry. However, several activities are unique to the storage of $CO_2$, namely, operations related to injecting $CO_2$ into geologic formations, as well as MVA for the $CO_2$ once it is injected. Under a broad range of scenarios for the deployment of carbon capture systems and the availability of $CO_2$ for either EOR operations or geologic storage, significant increases in EOR activity or geologic storage occur after 2025. Taking a national perspective, it appears that the industrial base to support pipeline development and EOR have the capacity to accommodate projected supplies of captured $CO_2$. If this activity is to become widely available commercially, however, the demonstration of geologic storage is critical, including both the ability to develop and operate sites at commercial scales safely and with minimal environmental disruption and the ability to ensure safe long-term storage.

## Findings

### Activities Supporting the $CO_2$ Storage Industrial Base Are Largely Shared with the Oil and Gas Sector

The $CO_2$ storage industrial base comprises three core activities: transportation of $CO_2$ by pipeline, EOR by $CO_2$ flooding, and geologic storage.

- *Pipelines.* The industrial base used to build and maintain natural gas and petroleum product pipelines is the same industrial base that would be used to build and maintain pipelines to transport $CO_2$. The same steel is used in pipelines in both industries. Pipeline construction techniques, and hence costs, are very similar. The major differences between pipelines used to transport $CO_2$ and those transporting natural gas and petroleum products concern the coatings and seals used for $CO_2$, the installation and operation of pumps needed to maintain pressure, and the presence of control valves to allow sections to be isolated for maintenance and to limit releases of $CO_2$ in case of a rupture. According to our analysis, the differences in costs between $CO_2$ pipeline equipment and equipment used in natural gas and petroleum product pipelines do not appreciably affect the ability of the industry to construct $CO_2$ pipelines.
- *$CO_2$–EOR.* Enhanced recovery by $CO_2$ flooding is already widely deployed commercially by the oil and gas industry. Oil companies survey, prepare sites, drill injection wells,

engage in well workovers, and plug wells used in EOR. Activities that are unique to EOR as opposed to other drilling operations include storing and injecting $CO_2$. Storage and injection involve receiving $CO_2$ from a bulk pipeline, distributing it throughout the field, injecting it into the field, and separating $CO_2$ from the produced crude oil.

- *Geologic storage.* Many activities supporting geologic storage are shared with the oil and gas sector, including geologic surveying, site preparation, and drilling wells. Injecting $CO_2$ is an activity shared with $CO_2$–EOR operations. Post-injection MVA operations must occur at geologic storage sites and at $CO_2$–EOR sites intending to demonstrate permanent storage. These activities are unique to carbon storage; the necessary technologies are being demonstrated but have not yet been deployed commercially.

### $CO_2$–EOR Can Facilitate the Development of Geologic Storage Industrial Capabilities

NETL, through the RCSPs, is demonstrating geologic storage of $CO_2$ and developing and testing technologies, systems, and protocols for carrying out MVA activities. From an equipment perspective, injecting $CO_2$ into a deep saline formation is similar to injecting $CO_2$ into a depleted oil reservoir. When $CO_2$–EOR is used for permanent storage, key supporting capabilities are developed. These supporting capabilities include detailed reservoir characterization, operational monitoring of the injected plume of $CO_2$, ensuring that $CO_2$ does not migrate into USDW, and long-term MVA activities.

Additional technologies need to be deployed to support geologic storage of $CO_2$. First, more subsurface mapping is needed because less is typically known about the geology of geologic storage sites than is known of EOR operations, which benefit from detailed records of the production history and geology of the field. Second, tracking and monitoring the $CO_2$ stream during injection will be different in geologic storage applications because there are no producing wells through which oil and $CO_2$ are recovered. Third, the quantity of $CO_2$ that would be injected into a single injection well is greater than that for a typical EOR injection well. When practiced for the purpose of carbon storage, $CO_2$–EOR advances industrial capabilities for carbon storage, but does not fully develop them.

### The Carbon Storage Industrial Base Has Largely Demonstrated the Capacity to Meet Development Needs for EOR and Geologic Storage

Because so much of the industrial base for EOR and $CO_2$ storage is the same or similar to that currently drawn upon for the natural gas and oil industries, we find no major barriers to ramping up operations to support $CO_2$ storage. In particular, we find:

- *The United States has already demonstrated the ability to lay likely needed lengths of pipelines for both EOR and CCS.* To support both EOR and deployment of carbon storage at the high end in the 2030–2035 timeframe, up to 32,000 miles of $CO_2$ pipelines would need to be constructed between 2025 and 2035 (roughly 3,200 miles per year). The United States has laid similar lengths of natural gas pipeline in the recent past. For example, the U.S. natural gas industry completed 3,600 miles of pipeline in 2008, and 21,000 miles between 2001 and 2010.
- *U.S. industry is likely to be able to hire sufficient workers with the skills needed to lay the potential length of pipeline needed to support both EOR and CCS.* The number of workers in the oil and gas pipeline construction industry grew by about 60 percent from 2005–2008, demonstrating the ability of the industry to quickly recruit and train labor during

periods of high demand. To meet the upper-bound estimate of $CO_2$ pipeline additions and provide a level of natural gas pipelines miles similar to the highest recent annual additions, the capacity of the pipeline construction industry would need to approximately double by 2025. Given the lead time available to build these pipelines and the likelihood that demand will actually be lower than this upper bound, the U.S. industrial base would likely have sufficient time to expand capacity to meet this demand.

- *We found no constraints on U.S. drilling capacity to expand EOR operations in our high-end EOR scenario.* On average, seven new EOR projects per year came on line from 2006 to 2010. We estimate that a maximum of 120 projects need to come online in the 2030–2035 timeframe, or 24 per year. In the context of the overall capabilities of the oil and gas sector, this constitutes a relatively small amount of activity. For example, we estimate the total number of drilling rigs required to support the highest pace of development to be 55, or slightly more than two active rigs per site. Currently, there are almost 2,000 onshore drilling rigs in operation in the United States; the number of rigs required to support EOR development would be a small fraction of the total.
- *We also found no constraints on the availability of drilling rigs or seismic crews to develop geologic storage in our high-end scenario.* Assuming that carbon capture systems are widely deployed soon and that the pace of deployment accelerates, 240 geologic storage sites may need to be opened in the five-year period from 2025–2030, an average of 48 sites per year, to accommodate growing volumes of $CO_2$. We estimate 84 drilling rigs would be required to open 48 sites per year. This number is a small fraction of the total onshore rigs currently available in the United States. We estimate that the number of active seismic survey teams needed to support this scale of development is approximately six, or one-tenth of the currently active teams today.

## Caveats and Limitations

This analysis considered the capabilities of the U.S. industrial base only in determining the ability to meet scenarios for EOR development and $CO_2$ availability. However, in the development of energy systems, technical capacity is only one consideration. We have omitted a key element of the broader CCS industrial base, namely the deployment of technologies to capture $CO_2$ from stationary sources. We assume that deployment of capture systems (as opposed to fuel switching) is the most economic compliance option and that systems to capture $CO_2$ from coal-fired power plants would be proven and deployed within a decade, with growing deployment thereafter at existing and new facilities. In the near term, the further development of U.S. natural gas resources might make large quantities of $CO_2$ available from natural gas processing. Both of these developments are highly uncertain. In this analysis, we have used scenarios to bound the ranges of potentially available $CO_2$; the scenarios themselves are estimates of the results of potential policies limiting emissions of $CO_2$. Whether such policies will be adopted is highly uncertain.

Similarly, we have assumed that the feasibility of long-term geologic storage would be demonstrated and available within the next decade.

This analysis also omitted regulatory and legal aspects of development of EOR or geologic storage. The process for obtaining the necessary permits for developing a geologic storage site takes at least two years. The development of interstate transmission pipeline systems would also be expected to require several years of design and permitting. Further, such processes can face delays or require redesign. Public perception (and that of policymakers) regarding carbon storage as a viable compliance option needs to be positive, thus providing support for deployment. If public support is not available, projects may be delayed or canceled, lessening the demands on the carbon storage industrial base.

We have not considered the potential for technological development. The technologies supporting pipeline operations, development and operations of EOR sites, and geologic storage continue to evolve and can be expected to perform better and cost less in the future. Geologic storage is in a period of demonstration across a range of geologies and scales. In most of our scenarios, large increases in the requirement to store $CO_2$, either through EOR operations or geologic storage, occur after 2025. Much technological progress can be made in the intervening 12 years.

Under most scenarios, we assume ten years before injection operations need to begin. Technological development, unforeseen natural events, changes in policy, and other factors will shape the future. One assumption underlying our analysis is that future core activities of the $CO_2$ storage industrial base will be similar to those of today. This is consistent with past experience in the pipeline and oil and gas sector, but the ability to construct pipelines or progress in EOR operations may decline. A substantial, long-term drop in oil prices would likely slow investment in EOR, reducing the ability of such operations to grow rapidly in the future and accommodate storage.

Our analysis has taken a national approach to characterizing the $CO_2$ storage industrial base. The U.S. $CO_2$ storage industrial base is largely capable of supporting the development of storage projects through the next 25 years. There are regional aspects, however, that we have ignored. For example, drilling rigs are distributed regionally according to oil and gas activity. While rigs are moved from site to site, interregional movements occur less frequently. Rigs could be in short supply during the development of initial geologic storage sites, which are likely to be geographically isolated from prior oil and gas activity.

## Implications for the NETL CCS Program

The NETL RCSPs are in the process of demonstrating geologic storage at commercial scale and in a range of geologies. The activities of the partnerships focus on the development of protocols for MVA of the stored carbon during and after $CO_2$ injection operations. Our analysis indicates that significant expansion of geologic storage capacity is required after 2025 under most of our scenarios. If several years are needed for permitting and siting, the United States has a window of approximately ten years before CCS needs to begin on a commercial scale. Based on the current activity of the partnerships, it appears from a technical perspective that the development of geologic storage is on track to meet this goal.

The industrial base for carbon transport and storage could be strained by demand for labor or equipment, much of which is shared with the oil and gas industrial base. During the RCSP demonstrations, NETL has the opportunity to collect data on project activity timelines and overall schedules, the number of qualified bidders, prices for critical equipment, and detailed labor costs. With these compiled data, and a comparison with external conditions in the oil and gas market, NETL will be able to ascertain whether the preliminary observed constraints on widespread deployment of carbon transportation and storage are likely to be binding, and determine appropriate and specific R&D strategies or recommended policy responses to alleviate these constraints.

APPENDIX A

# Listing of NAICS Codes and Occupational Codes

The North American Industrial Classification System (NAICS) catalogs economic activities. The NAICS codes of interest for this study are listed in Table A.1. These are the primary NAICS codes that support activities in the three major components of the $CO_2$ storage industrial base: pipeline transportation of $CO_2$, $CO_2$–EOR, and geologic storage.

For each NAICS code, the Bureau of Labor Statistics maintains data regarding the number and types of workers employed in that activity. These are occupation codes. Given the range of different skills required to support business operations, many of the occupation codes

**Table A.1**
**Industrial Classification Codes Relevant to the Industrial Base for $CO_2$ Storage**

| NAICS Code | NAICS Title | Pipeline Transportation of $CO_2$ | $CO_2$–EOR | Geologic Storage |
|---|---|---|---|---|
| 211111 | Crude Petroleum and Natural Gas Extraction | | X | |
| 213111 | Drilling Oil and Gas Wells | | X | X |
| 213112 | Support Activities for Oil and Gas Operations | | X | X |
| 237120 | Oil and Gas Pipeline and Related Structures Construction | X | | |
| 486110 | Pipeline Transportation of Crude Oil | X | | |
| 486210 | Pipeline Transportation of Natural Gas | X | | |
| 486910 | Pipeline Transportation of Refined Petroleum Products | X | | |
| 541360 | Geophysical Surveying and Mapping Services | | X | X |
| 541370 | Surveying and Mapping (except Geophysical) Services | | | |
| 333132 | Oil and Gas Field Machinery and Equipment Manufacturing | | X | X |
| 331210 | Iron and Steel Pipe and Tube Manufacturing from Purchased Steel | X | X | X |
| 332420 | Metal Tank (Heavy Gauge) Manufacturing | X | X | X |
| 333911 | Pump and Pumping Equipment Manufacturing | X | X | X |
| 333912 | Air and Gas Compressor Manufacturing | X | X | X |
| 532412 | Construction, Mining, and Forestry Machinery and Equipment Rental and Leasing | X | X | X |

SOURCE: RAND analysis.

are not necessarily critical (administrative services, for example). We have identified 37 occupation codes most relevant to the activities supporting $CO_2$ storage, listed below:

- 17–1022 Surveyors
- 17–2112 Industrial Engineers
- 17–2131 Materials Engineers
- 17–2141 Mechanical Engineers
- 17–2151 Mining and Geologic Engineers, Including Mining Safety Engineers
- 17–2171 Petroleum Engineers
- 17–3023 Electrical and Electronics Engineering Technicians
- 17–3024 Electromechanical Technicians
- 17–3026 Industrial Engineering Technicians
- 17–3027 Mechanical Engineering Technicians
- 17–3031 Surveying and Mapping Technicians
- 19–2041 Environmental Scientists and Specialists, Including Health
- 19–2042 Geoscientists, Except Hydrologists and Geographers
- 19–4041 Geologic and Petroleum Technicians
- 19–4091 Environmental Science and Protection Technicians, Including Health
- 29–9011 Occupational Health and Safety Specialists
- 29–9012 Occupational Health and Safety Technicians
- 47–0000 Construction and Extraction Occupations
- 47–2073 Operating Engineers and Other Construction Equipment Operators
- 47–2111 Electricians
- 47–2141 Painters, Construction and Maintenance
- 47–2152 Plumbers, Pipefitters, and Steamfitters
- 47–5011 Derrick Operators, Oil and Gas
- 47–5012 Rotary Drill Operators, Oil and Gas
- 47–5013 Service Unit Operators, Oil, Gas, and Mining
- 47–5021 Earth Drillers, Except Oil and Gas
- 47–5071 Roustabouts, Oil and Gas
- 47–5081 Helpers—Extraction Workers
- 47–5099 Extraction Workers, All Other
- 49–3042 Mobile Heavy Equipment Mechanics, Except Engines
- 49–9012 Control and Valve Installers and Repairers, Except Mechanical Door
- 51–4121 Welders, Cutters, Solderers, and Brazers
- 51–8092 Gas Plant Operators
- 51–8093 Petroleum Pump System Operators, Refinery Operators, and Gaugers.

APPENDIX B

# Listing of Top Firms by Revenue for Relevant NAICS Codes

This appendix lists the firms comprising the five-digit and six-digit NAICS codes identified as part of this analysis. It is important to note that firms self-report their membership in NAICS categories, therefore, there are differences between the five-digit and six-digit codes in terms of the firms represented. For example, several "major" oil and gas producers are listed in NAICS code 21111, but the six-digit code representing oil and gas extraction, 211111, contains oil and gas services firms.[1]

## Top Firms for Five-Digit NAICS Codes Relevant to the Carbon Storage Industrial Base

### Top Ten Firms by Revenue for Oil and Gas Extraction (NAICS Code 21111)

Exxon Mobil Corp.: $486.429 billion
Chevron Corp.: $253.706 billion
ConocoPhillips: $251.226 billion
Hess Corp.: $37.871 billion
Occidental Oil & Gas Corp.: $33.155 billion
Murphy Oil Corp.: $27.746 billion
Occidental Petroleum Corp.: $24.119 billion
Apache Corp.: $16.888 billion
FirstEnergy Corp.: $16.258 billion
Marathon Oil Corp.: $15.282 billion

### Top Ten Firms by Revenue for Support Activities for Mining (NAICS Code 21311)

Schlumberger Ltd.: $39.669 billion
Halliburton Co.: $24.829 billion
Baker Hughes Inc.: $19.831 billion
National Oilwell Varco, Inc.: $14.658 billion
Loews Corp.: $14.127 billion
Cameron International Corp.: $6.959 billion
FMC Technologies Inc.: $5.099 billion
MRC Global Inc.: $4.832 billion
Quanta Services Inc.: $4.624 billion
Chesapeake Operating, Inc.: $4.589 billion

[1] For this appendix, lists were generated using the LexisNexis company dossier/create-a-company-list function, plugging in each NAICS code for the United States. Revenues are reported for 2010 (LexisNexis, 2012).

**Top Ten Firms by Revenue for Oil and Gas Pipeline and Related Structures Construction (NAICS Code 23712)**

Targa Resources Corp.: $5.469 billion
Renewable Energy Group Inc.: $824 million
Wilbros Construction Us, LLC: $800 million
MYR Group Inc.: $780.400 million
Natco Group Inc.: $657.404 million
Integrated Electrical Services, Inc.: $475.363 million
C/O Willbros USA Inc.: $418 million
Lummus Technology Inc.: $363.4 million
Bechtel Oil, Gas and Chemicals, Inc.: $288.8 million
Northern Pipeline Construction Co.: $ 278.981 million

**Top Ten Firms by Revenue for Iron and Steel Pipe and Tube Manufacturing from Purchased Steel (NAICS Code 33121)**

KBR Inc.: $9.261 billion
Precision Castparts Corp.: $7.215 billion
McWane Inc.: $2 billion
Northwest Pipe Co.: $511.668 million
Handy and Harman Tube Co. Inc.: $496 million
WEBCO Industries Inc.: $465.648 million
Maverick Tube Corp.: $437.1 million
Allied Tube & Conduit Corp.: $380.8 million
Griffin Pipe Products Co. Inc.: $375 million
Wheatland Tube Co.: $340 million

**Top Ten Firms by Revenue for Metal Tank (Heavy Gauge) Manufacturing (NAICS Code 33242)**

Pall Corp.: $2.741 billion
Taylor-Wharton International LLC: $121.7 million
The Sterling Group LP: $86.2 million
Palmer Manufacturing & Tank Inc.: $79.579 million
Polar Tank Trailers, LLC: $73 million
CST Industries Inc.: $62.2 million
Ameri-Kart Michigan Corp,: $55.06 million
Tanco Engineering, Inc.: $48.535 million
Ameri-Kart Corp.: $46.904 million
TW Cylinders LLC: $32.5 million

**Top Ten Firms by Revenue for Mining and Oil and Gas Field Machinery Manufacturing (NAICS Code 33313)**

Joy Global Inc.: $4.404 billion
Bucyrus International Inc.: $3.651 billion
Smith International South America, Inc.: $3.600 billion
PSP Industries, Inc.: $3.493 billion
Baker Hughes Inc.: $3.141 billion
Smith International Inc.: $2.405 billion
Gardner Denver Inc.: $2.371 billion
Dresser-Rand Group Inc.: $2.312 billion
Weir SPM: $2.273 billion
IRI International Corp.: $1.700 billion

**Top Ten Firms by Revenue for Pump and Compressor Manufacturing (NAICS Code 33391)**

Envirotech Pumpsystems, Inc.: $8.538 billion
Dover Corp.: $4.446 billion
Flowserve Corp.: $4.032 billion
Pentair, Inc.: $3.031 billion
Beckett Corp.: $2.818 billion
ACD Co.: $2.572 billion
Exterran Holdings, Inc.: $2.462 billion
Goulds Pumps Inc.: $2.367 billion
Hypro Corp: $1.998 billion
Weir FloWay Inc.: $1.943 billion

**Top Ten Firms by Revenue for Pipeline Transportation of Crude Oil (NAICS Code 48611)**

Sunoco Logisitics Partners LP: $10.918 billion
Enbridge Energy Partners LP: $9.11 billion
Gulfmark Energy, Inc.: $2.006 billion
Magellan Midstream Partners LP: $1.749 billion
Alyeska Pipeline Service Co.: $658.007 million
Pacific Energy Partners LP: $224.302 million
Holly Energy Partners LP: $213.549 million
B P Oil Pipeline Co.: $153.700 million
TransMontaigne Partners LP: $152.292 million
Xi Capital Inc.: $150.700 million

**Top Ten Firms by Revenue for Pipeline Transportation of Natural Gas (NAICS Code 48621)**

Oneok Partners LP: $11.323 billion
Targa Resources Partners LP: $6.987 billion
Energy Transfer Partners LP: $6.850 billion
Williams Partners LP: $6.729 billion
Energy Transfer Equity, LP: $6.598 billion
Southwestern Energy Pipeline Co.: $2.146 billion
Crosstex Energy LP: $2.014 billion
PG&E Gas Transmission, Texas Corp.: $2 billion
Enogex LLC: $1.708 billion
Kinder Morgan Texas Pipeline LP: $1.7 billion

**Top Ten Firms by Revenue for Pipeline Transportation of Refined Petroleum Products (NAICS Code 48691)**

TEPPCO Partners LP: $13.533 billion
National Fuel Gas Supply Corp.: $1.770 billion
Magellan Midstream Holdings LP: $1.214 billion
Colonial Pipeline Co.: $824.064 million
Kaneb Pipe Line Operating Partnership LP: $640 million
Buckeye GP Holdings LP: $161.300 million
Plantation Pipe Line Co.: $130 million
Tampa Pipeline Corp.: $127.500 million
Nustar Pipeline Partners LP: $70.500 million
Te Products Pipeline Co., LP: $63.700 million

**Top Ten Firms by Revenue for Construction, Transportation, Mining, and Forestry Machinery and Equipment Rental and Leasing (NAICS Code 53241)**

United Rentals Inc.: $2.611 billion
Maxim Crane Works, LP: $1.978 billion
RSC Holdings Inc.: $1.234 billion
GATX Financial Corp.: $1.172 billion
Ashtead U.S. Holdings DGP: $627.2 million
Aircastle Limited: $527.710 million
TAL International Group Inc.: $516.687 million
Air Lease Corp.: $336.700 million
Exterran Energy Solutions, LP: $307 million
TTX Co.: $293 million

#### Top Ten Firms by Revenue for Geophysical Surveying and Mapping Services (NAICS Code 54136)

Meridian Oil Holding Inc.: $1.247 billion
Mariner Energy, Inc.: $942.941 million
Geokinetics Inc.: $763.729 million
Encore Operating LP: $696.527 million
Penn Virginia Resource Partners, LP: $656.704 million
Bois D'Arc Energy, Inc.: $355.460 million
Alsate Exploration, Inc.: $285 million
GHR Companies Inc.: $210.672 million
Ageons International Group Corp.: $157.110 million
GGI Liquidating Corp.: $105.523 million

#### Top Ten Firms by Revenue for Surveying and Mapping (Except Geophysical) Services (NAICS Code 54137)

Nolte Associates Inc.: $58 million
Carson Helicopters, Inc.: $24.5 million
Bury Plus Partners Inc.: $24.4 million
C.T. Consultants, Inc.: $21.5 million
Coler & Colantonio, Inc.: $17.5 million
American Surveying & Mapping, Inc.: $11 million
Davis, Bowen & Friedel, Inc.: $10.540 million
Anderson, Eckstein and Westrick, Inc.: $10 million
Consul-Tech Enterprises, Inc.: $9.588 million
Coast To Coast Survey Corp.: $8.8 million

### Top Firms for Six-Digit NAICs Codes Relevant to the Carbon Storage Industrial Base

#### Top Ten Firms by Revenue for Crude Petroleum and Natural Gas Extraction (NAICS Code 211111)

Occidental Oil and Gas Corp.: $33.155 billion
Schlumberger Ltd.: $28.931 billion
Petro-Canada: $27.743 billion
Murphy Oil Corporation: $23.345 billion
Occidental Petroleum Corp. of California: $18.16 billion
FirstEnergy Corp.: $13.339 billion
Williams Cos. Inc.: $9.616 billion
GDF Suez Energy International: $5.621 billion
Consol Energy Inc.: $5.236 billion
El Paso Corp.: $4.616 billion

**Top Ten Firms by Revenue for Drilling Oil and Gas Wells (NAICS Code 213111)**

Halliburton Co.: $17.973 billion
Loews Corp.: $14.615 billion
Pioneer Drilling Services, Ltd.: $7.798 billion
Chesapeake Operating, Inc.: $4.589 billion
Quanta Services Inc.: $3.931 billion
Diamond Offshore Drilling Inc.: $3.323 billion
Helmerich & Payne Inc.: $2.544 billion
Nabors Industries, Inc.: $2.142 billion
Rowan Cos. Inc.: $1.819 billion
Global Santa Fe Drilling Co.: $1.724 billion

**Top Ten Firms by Revenue for Support Activities for Oil and Gas Operations (NAICS Code 213112)**

Baker Hughes Inc.: $14.414 billion
National Oilwell Varco Inc.: $12.156 billion
FMC Technologies, Inc.: $4.126 billion
Chevron Investor Inc.: $3.18 billion
SEACOR Holdings Inc.: $2.649 billion
Exterran Holdings, Inc.: $2.462 billion
Oil States International Inc.: $2.412 billion
QEP Resources, Inc.: $2.246 billion
Hunt Consolidated, Inc.: $2.120 billion
Oceaneering International, Inc.: $1.917 billion

**Top Ten Firms by Revenue for Oil and Gas Pipeline and Related Structures Construction (NAICS Code 237120)**

Targa Resources Corp.: $5.469 billion
NATCO Group Inc.: $657.404 million
MYR Group Inc.: $597.1 million
Integrated Electrical Services, Inc.: $481.6 million
C/O Willbros USA Inc.: $418 million
Lummus Technology Inc.: $363.4 million
Bechtel Oil, Gas and Chemicals, Inc.: $288.8 million
MDU Construction Services Group, Inc.: $278.8 million
Basic Energy Services, LP: $259.7 million
Cajun Industries, LLC: $248.771 million

**Top Ten Firms by Revenue for Iron and Steel Pipe and Tube Manufacturing from Purchased Steel (NAICS Code 331210)**

KBR Inc.: $10.099 billion
McWane Inc.: $2 billion
ALRO Group: $973.304 million
WEBCO Industries Inc.: $465.648 million
Maverick Tube Corp.: $438.400 million
Northwest Pipe Co.: $386.750 million
Allied Tube & Conduit Corp.: $380.800 million
Griffin Pipe Products Co. Inc.: $375 million
Wheatland Tube Co.: $340 million
PTC Alliance Holdings Corp.: $289 million

**Top Ten Firms by Revenue for Metal Tank (Heavy Gauge) Manufacturing (NAICS Code 332420)**

Pall Corp.: $2.741 billion
Taylor-Wharton International LLC: $121.7 million
The Sterling Group LP: $86.2 million
Palmer Manufacturing & Tank Inc.: $79.579 million
Polar Tank Trailers, LLC: $73 million
CST Industries Inc.: $62.2 million
Tanco Engineering, Inc.: $48.535 million
Ameri-Kart Corp.: $46.904 million
TW Cylinders LLC: $32.5 million
NATGUN Corp.: $30.8 million

**Top Ten Firms by Revenue for Oil and Gas Field Machinery and Equipment Manufacturing (NAICS Code 333132)**

Cameron International Corp.: $6.135 billion
PSP Industries, Inc.: $3.493 billion
Special Projects Manufacturing Co.: $2.273 billion
Weatherford International, Inc.: $1.588 billion
Enviro Petroleum, Inc.: $1.5 billion
Fidelity Capital Investors, Inc.: $1.217 billion
Smith International Inc.: $1.127 billion
Stewart & Stevenson LLC: $861.2 million
Lufkin Industries Inc.: $645.643 million
Hydril Pressure Control: $503.048 million

### Top Ten Firms by Revenue for Pump and Pumping Equipment Manufacturing (NAICS Code 333911)

Envirotech Pumpsystems, Inc.: $8.538 billion
Dover Corp.: $4.446 billion
Flowserve Corporation: $4.032 billion
Pentair, Inc.: $3.031 billion
Goulds Pumps, Inc.: $2.367 billion
Hypro Corp.: $1.998 billion
Lincoln Automotive Co.: $368.142 million
Patterson Pump Co.: $296.808 million
Flowserve U.S. Inc.: $284.9 million
TD Group Holdings, LLC: $243.7 million

### Top Ten Firms by Revenue for Air and Gas Compressor Manufacturing (NAICS Code 333912)

Exterran Holdings Inc.: $947.707 million
Bristol Compressors Inc.: $483 million
Elliott Co.: $475.1 million
Arrow Pneumatics Co. Inc.: $404 million
Danfoss Scroll Technologies, LLC: $253.06 million
Dresser-Rand Holding (Delaware) LLC: $152.326 million
Sullair Corp.: $140 million
Cooper Turbocompressor Inc.: $130 million
Mitsubishi Heavy Industries Climate Control Inc.: $100 million
Bedford Precision Parts Corp.: $93.8 million

### Top Ten Firms by Revenue for Pipeline Transportation of Crude Oil (NAICS Code 486110)

Sunoco Logistics Partners LP: $7.838 billion
Magellan Midstream Partners LP: $1.558 billion
Alyeska Pipeline Service Co.: $658.007 million
Rose Rock Midstream, LP: $208.100 million
Holly Energy Partners LP: $182.097 million
B P Oil Pipeline Company: $153.700 million
TransMontaigne Partners LP: $150.899 million
Minnesota Pipe Line Company, LLC: $101.499 million
Enbridge (U.S.) Inc.: $62.6 million
Lubrication Services, LLC: $60 million

**Top Ten Firms by Revenue for Pipeline Transportation of Natural Gas (NAICS Code 486210)**

Kinder Morgan, Inc.: $11.846 billion
ONEOK Partners, LP: $8.676 billion
Energy Transfer Equity, LP: $6.598 billion
Williams Partners LP: $5.715 billion
Crosstex Energy Inc.: $1.793 billion
Enogex LLC: $1.708 billion
Kinder Morgan Texas Pipeline LP: $1.7 billion
Regency Energy Partners LP: $1.222 billion
Boardwalk Pipeline Partners LP: $1.117 billion
Columbia Gas Transmission, LLC: $1.02 billion

**Top Ten Firms by Revenue for Pipeline Transportation of Refined Petroleum Products (NAICS Code 486910)**

National Fuel Gas Supply Corp.: $1.77 billion
Magellan Midstream Partners LP: $1.320 billion
Colonial Pipeline Co.: $824.064 million
Kaneb Pipe Line Operating Partnership LP: $640 million
Explorer Pipeline Co.: $225.833 million
Buckeye GP Holdings LP: $161.3 million
Plantation Pipe Line Co.: $130 million
Tampa Pipeline Corp.: $127.5 million
Olympic Pipe Line Co.: $122 million
BP Corporation North America Inc.: $112.4 million

**Top Ten Firms by Revenue for Geophysical Surveying and Mapping Services (NAICS Code 541360)**

Penn Virginia Corp.: $673.864 million
Occidental Petroleum Investment Co. Inc.: $478.6 million
Alsate Exploration, Inc.: $285 million
Burlington Resources Oil & Gas Co.LP: $163.8 million
Schlumberger Technology Corp.: $105.7 million
GGI Liquidating Corp.: $105.523 million
Texaco Exploration India Inc.: $95.4 million
United Energy Corp.: $71.4 million
Rosbottom Interests, LLC: $57.1 million
First Energy Service Co.: $51.293 million

**Top Ten Firms by Revenue for Surveying and Mapping (Except Geophysical) Services (NAICS Code 541370)**

Nolte Associates Inc.: $58 million
Dof Subsea Usa, Inc.: $24.64 million
Carson Helicopters, Inc.: $24.5 million
Bury Plus Partners-Inc.: $24.4 million
C.T. Consultants, Inc.: $21.5 million
Coler & Colantonio, Inc.: $17.5 million
American Surveying & Mapping, Inc.: $11 million
Davis, Bowen & Friedel, Inc.: $10.54 million
Anderson, Eckstein and Westrick, Inc.: $10 million
Consul-Tech Enterprises, Inc.: $9.588 million

**Top Ten Firms by Revenue for Construction, Mining, and Forestry Machinery and Equipment Rental and Leasing (NAICS Code 532412)**

Maxim Crane Works, LP: $1.978 billion
Ashtead US Holdings DGP: $627.2 million
Prime Service, Inc.: $500 million
RSC Holdings II, LLC: $474.8 million
Exterran Energy Solutions, LP: $307 million
Ahern Rentals, Inc. : $284.321 million
Ashtead Holdings, LLC: $214.5 million
Exterran, Inc.: $178.2 million
Kirby-Smith Machinery, Inc.: $149.331 million
Cecil I. Walker Machinery Co.: $93.5 million

## References


Advanced Resources International, *Basin Oriented Strategies for $CO_2$ Enhanced Oil Recovery: Permian Basin*, Arlington, Va.: February, 2006.

American Association of Professional Landmen, *About AAPL*, 2012. As of December 20, 2012: http://www.landman.org/about-aapl

"American Carbon Capture and Storage Industry Starts Capturing $CO_2$," Environment News Service, 2011. As of December 18, 2012:
http://ens-newswire.com/2011/06/22/american-carbon-capture-and-storage-industry-starts-capturing-co2/

Berkman, Steve, and Tory Stokes, "National Oilwell Varco, 58th Annual Rig Census," *World Oil*, November 2011, pp. 70–79.

Big Sky Carbon Sequestration Partnership, *Schedule*, Missoula, Mont.: University of Montana, undated.

Blinclow, Mike, "Denbury: EOR and $CO_2$ Supply Update," *17th Annual $CO_2$ Flooding Conference*, Midland, Texas, 2011.

Bliss, Kevin, Darrick Eugene, Robert W. Harms, Victor G. Carrillo, Kipp Coddington, Mike Moore, John Harju, Melanie Jensen, Lisa Botnen, Philip M. Marston, Doug Louis, Steve Melzer, Colby Drechsel, Jack Moody, and Lou Whitman, *A Policy, Legal and Regulatory Evaluation of the Feasibility of National Pipeline Infrastructure for the Transport and Storage of Carbon Dioxide*, Norcross, Georgia: Southern States Energy Board, September 10, 2010.

BLM—*See* Bureau of Land Management.

Brown, Daryl, Jim Cabe, and Tyson Stout, "National Lab Uses OGJ Data to Develop Cost Equations," *Oil & Gas Journal*, Vol. 109, No. 1, 2011, p. 108.

Bureau of Labor Statistics, "NAICS-Based Quarterly Census of Employment and Wages (QCEW) Data," 2012.

Bureau of Land Management, *Greencore Pipeline Company LLC Environmental Assessment*, 2011. As of December 20, 2012:
http://www.blm.gov/pgdata/content/wy/en/info/NEPA/documents/cfo/greencore.html

California General Assembly, *California Global Warming Solutions Act of 2006*, Assembly Bill 32. As of December 20, 2012:
http://www.leginfo.ca.gov/pub/05-06/bill/asm/ab_0001-0050/ab_32_bill_20060927_chaptered.pdf

CFR—*See* Code of Federal Regulations.

Chandel, Munish K., Lincoln F. Pratson, and Eric Williams, "Potential Economies of Scale in $CO_2$ Transport Through Use of a Trunk Pipeline," *Energy Conversion and Management*, Vol. 51, No. 12, 2010, pp. 2825–2834.

Code of Federal Regulations, Title 40, Underground Injection Control Program, Part 146, Criteria and Standards, 2011a. As of December 18, 2012:
http://www.law.cornell.edu/cfr/text/40/146

———, Title 49, Part 195, Subpart D-Construction, 2011b. As of December 18, 2012:
http://www.law.cornell.edu/cfr/text/49/195/subpart-D

———, Title 49, Part 192, Subpart E-Welding of Steel in Pipelines, 2011c. As of December 18, 2012:
http://www.law.cornell.edu/cfr/text/49/192/subpart-E

Davis, Darrell, Mark Scott, Adam Robinson, Kris Roberson, and Bill Freeman, "Large Scale $CO_2$ Flood Begins Along Texas Gulf Coast: Technical Challenges in Reactivating an Old Oil Field," *17th Annual $CO_2$ Flooding Conference*, Midland, Texas, 2011.

Det Norske Veritas, *Recommended Practice DNV–RP–J202: Design and Operation of $CO_2$ Pipelines*, April 2010.

DNV—*See* Det Norske Veritas.

Dooley, James J., Robert T. Dahowski, and Casie L. Davidson, "Comparing Existing Pipeline Networks with the Potential Scale of Future U.S. $CO_2$ Pipeline Networks," *Energy Procedia*, Vol. 1, No. 1, 2009, pp. 1595–1602. As of December 20, 2012:
http://www.sciencedirect.com/science/article/pii/S1876610209002100

EIA—*See* U.S. Energy Information Administration.

Element Energy Limited, *$CO_2$ Pipeline Infrastructure*, Cambridge, UK: International Energy Agency Greenhouse Gas Programme, April 27, 2010.

EPA—*See* U.S. Environmental Protection Agency.

Folga, Stephan M., *Natural Gas Pipeline Technology Overview*, Argonne, Ill.: Argonne National Laboratory, ANL/EVA/TM/08–5, 2007.

Fugleberg, Jeremy, "Construction of Carbon Dioxide Pipeline Starts This Year," *Casper Star Tribune*, May 26, 2011. As of December 20, 2012:
http://trib.com/business/energy/article_bed8ad33-83e9-5861-ba12-50dc2fb69cb0.html

Gale, John, and John Davison, "Transmission of $CO_2$—Safety and Economic Considerations," *Energy*, Vol. 29, No. 9, 2004, pp. 1319–1328.

Grant, Timothy, "Information Regarding MVA and Timelines for MGSC and SECARB Tests," correspondence to David Ortiz, RAND, Pittsburgh, Pa., November 1, 2012.

ICF International, *Developing a Pipeline Infrastructure for $CO_2$ Capture and Storage: Issues and Challenges*, Washington, D.C.: Interstate Natural Gas Association of America, February 2009.

Interliance Consulting, Inc., *Critical Skills Forecast for the Natural Gas Transmission Industry*, INGAA Foundation, 2009. As of January 4, 2013:
http://www.ingaa.org/File.aspx?id=12854

Interstate Natural Gas Association of America, *Line Pipe*, 2012. As of December 20, 2012:
http://www.ingaa.org/cms/112.aspx

Koperna, George, David Riestenberg, Robin Petrusak, Richard Esposito, and Dick Rhudy, "Lessons Learned While Conducting Drilling and $CO_2$ Injection Operations at the Victor J. Daniel Power Plant in Mississippi," paper presented at 2009 Society of Petroleum Engineers Annual Technical Conference and Exhibition, New Orleans, La., Society of Petroleum Engineers, October 4–7, 2009.

Kuby, Michael J., Richard S. Middleton, and Jeffrey M. Bielicki, "Analysis of Cost Savings from Networking Pipelines in CCS Infrastructure Systems," *Energy Procedia*, Vol. 4, 2011, pp. 2808–2815. As of December 20, 2012:
http://www.sciencedirect.com/science/article/pii/S1876610211003821

LexisNexis, company dossiers list function, 2012. As of January 9, 2013:
http://www.lexisnexis.com/hottopics/lnacademic/

Liu, Hengwei W., and Kelly Sims Gallagher, "Preparing to Ramp Up Large-Scale CCS Demonstrations: An Engineering-Economic Assessment of $CO_2$ Pipeline Transportation in China," *International Journal of Greenhouse Gas Control*, Vol. 5, No. 4, July 2011, pp. 798–804. As of December 20, 2012:
http://www.sciencedirect.com/science/article/pii/S1750583610001714

Massachusetts Institute of Technology Energy Initiative, *Non-Power Plant Carbon Dioxide Capture and Storage Projects*, March 17, 2012. As of December 20, 2012:
http://sequestration.mit.edu/tools/projects/storage_only.html

McCoy, Sean T., and Edward S. Rubin, "An Engineering-Economic Model of Pipeline Transport of $CO_2$ with Application to Carbon Capture and Storage," *International Journal of Greenhouse Gas Control*, Vol. 2, No. 2, 2008, pp. 219–229. As of December 20, 2012:
http://www.sciencedirect.com/science/article/pii/S1750583607001193

Melzer, Steven, "$CO_2$ EOR Project Timeline," paper presented at *17th Annual $CO_2$ Flooding Conference*, Midland, Texas, December 2011.

Midwest Geological Sequestration Consortium, "Deep Saline Storage, Illinois Basin—Decatur Project," 2012. As of December 20, 2012:
http://sequestration.org/mgscprojects/deepsalinestorage.html

Morgan, M. Granger, and Sean T. McCoy, *Carbon Capture and Sequestration: Removing the Legal and Regulatory Barriers*, New York, NY: Resources for the Future Press, Taylor and Francis, 2012.

Moritis, Guntis, "$CO_2$ Miscible, Steam Dominate Enhanced Oil Recovery Processes," *Oil and Gas Journal*, April 19, 2010.

National Energy Technology Laboratory, *Fact Sheet for Partnership Field Validation Test*, Decatur, Illinois, undated. As of December 20, 2012:
http://www.netl.doe.gov/publications/proceedings/07/rcsp/factsheets/20-MGSC_Illinois Basin Saline Formation Test.pdf

———, *Best Practices for Monitoring, Verification, and Accounting of $CO_2$ Stored in Deep Geologic Formations*, Pittsburgh, Pa., Department of Energy/NETL–311/081508, January 2009.

———, *Best Practices for Site Screening, Site Selection, and Initial Characterization for Storage of $CO_2$ in Deep Geologic Formations*, Pittsburgh, Pa.: Department of Energy/NETL–401/090808, November, 2010.

———, *Improving Domestic Energy Security and Lowering $CO_2$ Emissions with Next Generation $CO_2$-Enhanced Oil Recovery ($CO_2$–EOR)*, Pittsburgh, Pa., June 20, 2011a.

———, *Statement of Project Objectives: Geological Characterization of the South Georgia Rift Basin for Source Proximal $CO_2$ Storage*, Pittsburgh, Pa., November 21, 2011b.

———, *Statement of Project Objectives: Modeling $CO_2$ Sequestration in Saline Aquifer and Depleted Oil Reservoir to Evaluate Regional $CO_2$ Sequestration Potential of the Ozark Plateau Aquifer System, South-Central Kansas*, Pittsburgh, Pa., 2011c.

———, *Carbon Sequestration: Geologic Storage Focus Area*, 2012a. As of December 20, 2012:
http://www.netl.doe.gov/technologies/carbon_seq/corerd/storage.html

———, *Carbon Sequestration: Monitoring, Verification, and Accounting (MVA) Focus Area*, 2012b. As of December 20, 2012:
http://www.netl.doe.gov/technologies/carbon_seq/corerd/mva.html

———, *Carbon Sequestration: RCSP Geologic Characterization Efforts*, 2012c. As of December 20, 2012:
http://www.netl.doe.gov/technologies/carbon_seq/infrastructure/charefforts.html

———, *Carbon Sequestration: Regional Partnership Development Phase (Phase III) Projects*, 2012d. As of December 20, 2012:
http://www.netl.doe.gov/technologies/carbon_seq/infrastructure/rcspiii.html

———, *$CO_2$ Storage Cost Model*, Pittsburgh, Pa., February 27, 2012e.

NETL—*See* National Energy Technology Laboratory.

Occidental of Elk Hills, Inc., *Oxy Elk Hills $CO_2$–EOR Project: MRV Plan Draft*, Houston, Texas, July 23, 2010.

PHMSA—*See* Pipeline and Hazardous Materials Safety Administration.

Pipeline and Hazardous Materials Safety Administration, *2010 Hazardous Liquid Annuals Data*, 2012a. As of December 20, 2012:
https://explore.data.gov/Transportation/Pipeline-Annual-Data-2010-Hazardous-Liquid-Annuals/6pbj-j2af

———, *Natural Gas Transmission, Gas Distribution, and Hazardous Liquid Pipeline Annual Mileage*, 2012b. As of December 20, 2012:
http://phmsa.dot.gov/portal/site/PHMSA/menuitem.ebdc7a8a7e39f2e55cf2031050248a0c/?vgnextoid=036b52edc3c3e110VgnVCM1000001ecb7898RCRD&vgnextchannel=3b6c03347e4d8210VgnVCM1000001ecb7898RCRD&vgnextfmt=print

Samaras, Constantine, Jeffrey A. Drezner, Henry H. Willis, and Evan Bloom, *Characterizing the U.S. Industrial Base for Coal-Powered Electricity*, Santa Monica, Calif.: RAND Corporation, MG-1147-NETL, 2011. As of December 20, 2012:
http://www.rand.org/pubs/monographs/MG1147.html

Schoots, Koen, Rodrigo Rivera-Tinoco, Geert Verbong, and Bob van der Zwaan, "Historical Variation in the Capital Costs of Natural Gas, Carbon Dioxide and Hydrogen Pipelines and Implications for Future Infrastructure," *International Journal of Greenhouse Gas Control*, Vol. 5, No. 6, 2011, pp. 1614–1623. As of December 20, 2012:
http://www.sciencedirect.com/science/article/pii/S1750583611001782

Seong, Somi, Obaid Younossi, Benjamin W. Goldsmith, Thomas Lang, and Michael J. Neumann, *Titanium: Industrial Base, Price Trends, and Technology Initiatives*, Santa Monica, Calif.: RAND Corporation, MG-789-AF, 2009. As of December 20, 2012:
http://www.rand.org/pubs/monographs/MG789.html

Southeast Regional Carbon Sequestration Partnership, *Gulf Coast Stacked Storage Project*, Norcross, Georgia, undated. As of December 20, 2012:
http://www.secarbon.org/files/gulf-coast-stacked-storage-project.pdf

Toman, Michael, Aimee E. Curtright, David S. Ortiz, Joel Darmstadter, and Brian Shannon, *Unconventional Fossil-Based Fuels: Economic and Environmental Trade-Offs*, Santa Monica, Calif.: RAND Corporation, TR-580-NCEP, 2008. As of December 20, 2012:
http://www.rand.org/pubs/technical_reports/TR580.html

United Association, *Apprenticeship*, 2011. As of January 3, 2013:
http://www.ua.org/apprenticeship.asp

U.S. Energy Information Adminstration, "About U.S. Natural Gas Pipelines," undated. As of December 20, 2012:
http://www.eia.gov/pub/oil_gas/natural_gas/analysis_publications/ngpipeline/develop.html

———, "Crude Oil and Natural Gas Drilling Activity," March 2, 2012. As of December 20, 2012:
http://www.eia.gov/dnav/pet/pet_crd_drill_s1_m.htm

U.S. Environmental Protection Agency, *Cost Analysis for the Federal Requirements Under the Underground Injection Control Program for Carbon Dioxide Geologic Sequestration Wells (Final GS Rule)*, Washington, D.C.: Office of Water, EPA 816–R10–013, 2010.

———, "Carbon Pollution Standard for New Power Plants: Regulatory Actions," March 27, 2012. As of December 20, 2012:
http://www.epa.gov/airquality/cps/actions.html

U.S. Government Accountability Office, *Coal Power Plants: Opportunities Exist for DOE to Provide Better Information on the Maturity of Key Technology to Reduce Carbon Dioxide Emissions*, Washington, D.C., June, 2010.

U.S. House of Representatives, *American Clean Energy and Security Act*, H.R. 2454, 2009.

Van der Zwaan, Bob C. C., Koen Schoots, Rodrigo Rivera-Tinoco, and Geert P. J. Verbong, "The Cost of Pipelining Climate Change Mitigation: An Overview of the Economics of $CH_4$, $CO_2$ and $H_2$ Transportation," *Applied Energy*, Vol. 88, No. 11, 2011, pp. 3821–3831. As of December 20, 2012:
http://www.sciencedirect.com/science/article/pii/S030626191100314X

Wehner, Scott, "U.S. $CO_2$ and $CO_2$–EOR Developments," *17th Annual $CO_2$ Flooding Conference*, Midland, Texas, 2011.